PHILIP'S

MOON
OBSERVER'S GUIDE

Peter Grego

For Tina and Jacy – my two bright lunar rays

PETER GREGO has regularly observed the skies since 1976, and commenced studying the Moon in 1982. Based in Cornwall, he observes with a variety of instruments, ranging from a vintage 75 mm refractor to a home-made 300 mm Newtonian. His primary interests are the Moon and bright planets, but he occasionally likes to 'go deep' when the Moon isn't up.

Grego has been Director of the Lunar Section of Britain's Society for Popular Astronomy (SPA) since 1984 and is the Assistant Director of the British Astronomical Association (BAA) Lunar Section. He edits a number of astronomy publications, including *Luna* (journal of the SPA Lunar Section), the *BAA Lunar Section Circular*, *The New Moon* (topographical journal of the BAA Lunar Section), the *SPA News Circulars* and *Popular Astronomy* magazine, and is layout editor of the *Bulletin* of the Society for the History of Astronomy. In addition, Grego writes and illustrates the monthly MoonWatch page in *Astronomy Now* magazine and is the expert on observing for *Sky at Night* magazine's Astro Answers. He is a member of the British Astronomical Association and a Fellow of the Royal Astronomical Society, and is the author of 16 astronomy books.

First published in Great Britain in 2003 by Philip's,
a division of Octopus Publishing Group Limited
(www.octopusbooks.co.uk)
Endeavour House, 189 Shaftesbury Avenue,
London WC2H 8JY
An Hachette UK Company (www.hachette.co.uk)

This revised edition 2010
Reprinted 2010

ISBN 978-1-84907-065-2

Printed in China

Details of other Philip's titles and services can be found on our website at: **www.philips-maps.co.uk**

Title page: The 16-day-old Moon photographed on 31 December 2001 by the author, using a 250 mm reflector with a Ricoh RDC-5000 digicam.

CONTENTS

── *WHY OBSERVE THE MOON?* ──

*From all time the Moon has had the privilege of charming the gaze,
and attracting the particular attention of mortals. What thoughts have
not risen to her pale, yet luminous disk? Orb of mystery and of solitude,
brooding over our silent nights, this celestial luminary is at once sad
and splendid in her glacial purity, and her limpid rays provoke
a reverie full of charm and melancholy.*
Camille Flammarion, *Astronomy for Amateurs*, 1903

An evening twilight sky adorned with a gleaming crescent Moon is one of nature's most sublime spectacles. Our celestial neighbour has inspired poets and artists down the centuries, and though astronomers may know the Moon intimately these days, the lunar magic experienced by philosophers of old remains as powerful as ever.

Sunset, 15 May 2002. At the foot of the hill, my 15 kg rucksack had felt comfortable on my back. Inside it is my trusty Maksutov telescope with its computer-driven altazimuth mount and lightweight aluminium tripod. Now, halfway up the hill, this 'portable' instrument is growing heavy. Pausing to catch my breath, I turn to the darkening eastern sky. Rising towards the zenith is the deep purple swathe of the Earth's shadow – the same shadow that tracks across the face of the Moon at a lunar eclipse. Several pale stellar flecks are now visible. Arcturus, the brightest of them, twinkles with orange starlight – which, I muse, set out in 1965, the year I was born.

I continue up the slope towards the darkening orange-green sunset. As I near the brow of the hill, the distant cityscape of Birmingham comes into view – myriad streetlights amid dimming patterns of roads and buildings. Now on level ground, I round the line of oaks and sycamores to my left and scan the deep orange afterglow that hugs the sunset horizon. The first in tonight's grand planetary line-up, faint little Mercury, emerges from behind a tree trunk, just clear of the north-western horizon. Saturn springs into view, higher in the sky, followed by a dim ruddy Mars and a truly dazzling white Venus. Finally, I catch sight of Jupiter and the crescent Moon, side by side in Gemini, their light playing through the leaves of an old oak. There's nobody around to hear my gasp of delight. Now at my observing site, I survey the whole splendid planetary array – an alignment whose rarity has spurred me to the top of Beacon Hill.

After setting up the telescope, which takes no more than five minutes, I gaze briefly at each planet though the eyepiece. For all the glamour of the planets, only one object can hold my fascinated gaze for more than a few moments – the Moon. For the next couple of hours it will be mine to marvel at and explore. An observational drawing makes a fine record of the evening's viewing, but it also requires discipline and a certain degree of objective detachment. Tonight, I am enthralled by the shadow-play along the edge of the lunar crescent. My sketchpad and pencils will remain untouched this evening. I'm happy to be a bedazzled tourist now and again.

I stare at the Moon through a low-power eyepiece, taking in the whole gleaming crescent in the same field of view. I can see the faint, cool blue glow of earthshine on the dark side of the Moon. Beneath the prominent dark oval of the Sea of Crises near the eastern limb is the crater Cleomedes. As soon as I catch sight of it, my mind is transported back exactly 19 years and a day, when the Moon was a crescent much like tonight, also in Gemini but then accompanied by Venus. Cleomedes was the first lunar crater I'd ever drawn. By any standards it was a clumsy effort – in two hours of observing I reduced a remarkable crater to a wholly unremarkable sketch. Yet it had been a start, and I'd become hooked.

Since that time, through patient observing and recording, the lunar landscape has become to me a broadly familiar place, yet always full of wonder. Tonight only a sliver of Moon is visible, and the eastern lunar seas and their surrounding craters provide a visual delight until the Moon sinks into the haze above the city and its image dims, shimmers and degrades.

The Moon appears so big, bright and full of detail that the smallest optical instruments are capable of revealing more of its wonders. Through binoculars the crescent Moon is magnified into a sickle with

▶ An evening scene with the crescent Moon, Jupiter (right), Mars (farther right) and Saturn (top), photographed on 6 April 2000 by Paul Stephens.

a dented edge, while a telescope will transform it into a spectacular alien landscape. Most people who look at the Moon through the telescope do so for the pure visual pleasure it brings – in its own way, an activity just as meaningful as the highest level of scientific enquiry. Anyone can admire the Moon and enjoy it at face value. Equally, everyone is at liberty to find out more about it.

Ever since Galileo first sketched the lunar craters, observers have wanted to keep a permanent record of their travels around the Moon's surface by sketching what they see through the eyepiece. Drawing the lunar landscape might seem an eccentric and outmoded pursuit, now that detailed images obtained by professional observatories and spacecraft are readily available. But lunar observers who take the trouble to sketch what they see will discover an immensely useful and rewarding activity which will improve every single aspect of their observing skills. It is the only way to thoroughly learn your way around the Moon – looking at photographs is no substitute.

As you draw what you see through a telescope, your eye is being trained to register small details that the untrained eye might overlook. The Moon's surface is packed with very fine detail, and your ability as an observer to discern it will continually improve as you spend more hours at the eyepiece. With practice comes confidence and skill. The discipline of making accurate lunar drawings, learned at the eyepiece, will pay high dividends in other fields of amateur astronomy, such as planetary observation. If you can master the art of drawing lunar features, the planets and deep-sky objects will be easy to depict!

During the course of your lunar apprenticeship, the apparent confusion of the Moon's landscape, with its arcane nomenclature, gives way to an increasing familiarity. Once you have learned to record lunar features, you can begin to pursue programmes of observation in earnest. These may be driven by a particular interest – a passion for rayed craters, say – or you can follow an observing programme organized by an astronomical society.

Real discoveries can still be made by the skilled amateur lunar observer. Uncharted, unsuspected small-scale features can show themselves under very low angles of illumination. Lunar topography is virtually neglected by professional astronomers, so it's up to amateurs to keep the Moon under scrutiny. And it's possible that at some point in the future a small asteroid will carve out a new crater on the Moon large enough to be seen from the Earth. More than likely an amateur astronomer would be the first to detect such an impact and view the hot new blemish on the Moon's face.

Drawing is one specialized aspect of lunar observation. The Moon's surface can also of course be recorded in great detail by conventional photography, or by digital imaging using a digital camera, modified webcam, dedicated astronomical CCD camera or analogue/digital videography. Learning how to master the appropriate techniques – both at the eyepiece and in image processing at the computer – can be just as demanding as learning how to observe and draw lunar features. Other, more exotic means of recording the Moon's surface include infrared imaging and spectroscopy.

Of course, all observations – whether in the form of topographical sketches, CCD images or advanced work in other wavelengths – provide the raw materials for further research. Nobody who observes the Moon can fail to formulate their own ideas about how various topographic features formed, and even today there is much to be learned about the precise sequence of events that sculpted the lunar landscape. Fresh perspectives on familiar features may be gained by enhancing and manipulating images on computer, giving new insights into the Moon. Enhancing colour saturation on images can reveal distinct differences in lunar mineralogy, revealing the layering of mare lavas. Computer programs capable of rendering views of features from any desired angle enable near-limb features to be seen in a fresh light, and computers may be used to create 3D anaglyphs of lunar terrain, adding a new dimension to research.

As well as topographic studies, there are a number of exciting avenues to explore in the fleeting lunar phenomena that are occasionally visible. So-called transient lunar phenomena are apparent obscurations, glows or flashes on the Moon's surface which, because of their extreme rarity, observers must be vigilant and skilled to detect and record successfully. Occasionally the Moon passes in front of bright stars and planets. Such lunar occultations are a joy to observe and provide wonderful photo-opportunities. Across the world, thousands of amateurs routinely observe and time occultations of fainter stars. These timings are valuable scientific data, since they help in fine-tuning the parameters of the Moon's orbit and the rate of the Earth's rotation.

Bright occultations and solar and lunar eclipses make us more aware that the Moon is capable of producing some of nature's most spectacular sights. Equally as spectacular, the lunar landscape is accessible through the most modest optical equipment. Lunar observing is by far the most visually rewarding branch of amateur astronomy, and it's no wonder that for centuries the Moon has attracted some of the world's greatest telescopic observers. Our satellite's rugged surface provides the greatest show off Earth.

Peter Grego
Cornwall, UK
September 2009

Note: *All images in the book, whether photographs, CCD images, maps or drawings, are oriented with north upwards. This is the naked-eye or binocular view from the northern hemisphere, and the telescopic view from the southern hemisphere.*

LUNAR GEOLOGY AND
THE MOON'S FEATURES

An immensely long and chequered geological history is portrayed in the face of the Moon. Set in solid rock and clearly visible through the telescope eyepiece are the results of a few billion years of asteroidal bombardment, volcanism and crustal activity. Few solid worlds in the Solar System display such clearly chiselled features chronicling such a long period of change. The Earth, Mars and Venus have had their topographies modified by recent volcanic activity in conjunction with varying degrees of crustal movement and surface erosion. Scars of the more ancient asteroidal impacts have been scoured from their surfaces, buried by sediment and deformed by crustal activity. But a strikingly different story is told by the lunar surface. The Moon has never possessed a substantial atmosphere to power various forms of surface erosion; neither has it retained a thin, mobile crust that deforms to the deep-seated whims of a dynamic molten interior. It is a sobering fact that many lunar features visible through the telescope are older than the most ancient rocks on Earth – indeed, older than terrestrial life itself.

At first glance, the cause of some of the Moon's features seems clear. The lunar rays, for instance, look like lines or sheets of material thrown out from craters at the time of their formation – and that is what they are. However, the origin of other features is not so easy to deduce: what, for example, could have caused the mysterious rounded hills known as domes, or the sinuous wrinkle ridges that wind their way across the lunar maria? Until the advent of lunar exploration and the return of lunar samples for study on Earth, the problem with lunar geology (or 'selenology', to use an older term) was that the appearance of the Moon's topography – dramatic and detailed though it is – only represents half the picture. It is like trying to interpret Rembrandt's painting techniques from a photograph of a painting rather than by examining the work directly. Yet this lack of hard, hands-on evidence did not deter generations of selenologists from formulating their own theories and expounding them in print. It took the space age, with detailed imaging of the Moon by orbiting spacecraft and examination of lunar rocks returned by US and Russian missions, to open up the Moon's secrets.

The Moon's craters

A lunar crater may be defined as any circular or near-circular indentation in the Moon's surface. Its floor does not have to be depressed very far below the level of the surrounding terrain, nor must it be strongly bowl-shaped in profile for it to qualify as a crater. Paradoxically, flooding with lava has given a few craters floors which are raised above the mean level of the surrounding terrain, the crater-plateau Wargentin being the most famous example of its kind. Lunar geologists prefer to call the larger features (ones in excess of 300 km in diameter) ringed basins rather than craters.

Most people picture a lunar landscape as a starkly illuminated grey vista packed full of craters. It is estimated that the Moon's surface is studded with more than 3 trillion (3,000,000,000,000) craters larger than a metre in diameter. The vast majority of these are small pits hewn out by the impact of small meteoroids. Craters larger than around 3 km in diameter are comfortably within the resolution of the average 150 mm aperture telescope, and there are more than 10,000 of

▲ The Moon, showing a waning gibbous phase. Visible are the western lunar seas, the crater-crowded southern highlands and four bright ray craters, Aristarchus, Copernicus, Kepler and Tycho. The image was obtained on 10 August 2000 by Peter Grego, using a 150 mm refractor with a Ricoh RDC-5000 digicam.

these visible on the nearside. Some areas, like the maria, appear relatively smooth and crater-free, while others, like the southern highlands, are gridlocked with craters.

Cratering by internal forces

The few remaining supporters of endogenic lunar cratering theories – in reality, only a small minority of mainly amateur astronomers – take elements from old, long-discredited volcanic and plutonic theories in order to grant the Moon a cratering heritage fuelled by internal forces. One of the most popular endogenic cratering theories sees the development of a large lunar dome resulting from an upwelling of magma known as a laccolith. Faulting then occurs around the edges of this dome, and subsidence produces a crater that may (or may not) be prone to lava flooding. Such features do exist on Earth and on Venus, but they are not very common. Moreover, laccolith collapses cannot account for the complexity of most of the lunar craters, including their central mountains, terraced walls, and surrounding ridges and rays.

Craters with impact

Much of the more recent work on impact crater theory was carried out by the late American geologist Eugene Shoemaker, who founded the Astrogeology Branch of the US Geological Survey (USGS) in 1961. Shoemaker – probably more famous for his co-discovery of Comet Shoemaker–Levy 9 which impacted on Jupiter in 1994 – set the USGS on a unique project to map lunar geology using detailed photographs, telescopic observations and spacecraft images. This culminated in 1971 with the publication of the 1:5,000,000 scale *Geologic Map of the Near Side of the Moon* by Jack McCauley and Don Wilhelms, which showed the relationship between distinct geological units, from mare lava fills to ancient, heavily cratered lunar crust. It was the last, greatest lunar geological map based on pre-Apollo data.

These days we possess a lot of hard scientific information about the make-up of the Moon's surface, derived from analysis of lunar rock samples and close-up images taken from lunar orbit and on the lunar surface. Some proponents of volcanism have suggested that Apollo scientists were biased in favour of the impact theory of cratering, and that the results were interpreted to fit the theory. In reality, the Apollo science teams pored over the evidence with nothing less than open minds. The explorations were conducted with a view to gathering as much scientific information as possible, and with each new mission came increasing scientific rigour. Indeed, the last Apollo mission took Harrison Schmitt, a qualified geologist, to the valley of Taurus-Littrow.

The Apollo missions sampled six geologically diverse locations, chosen as being representative of the nearside terrain, and covering many tens of square kilometres. Several international teams were given the opportunity to analyse the data obtained on the lunar surface and make studies of the lunar rock and soil samples. An enormous amount of physical evidence was obtained by Apollo, including the return to Earth of 379 kg of lunar rock and soil. This, and the observations made from lunar orbit by the Clementine (1994) and Lunar Prospector (1998) spacecraft, yielded very little to support the idea that volcanism formed any of the Moon's major craters.

The Moon's craters, great and small

All the Moon's ringed basins, walled plains and the overwhelming majority of craters visible through the telescope were formed by asteroidal impact. Despite this, no two are exactly alike. Even craters of similar size on similar terrain can look utterly different from each other, though their mode of formation was identical.

Most of the Moon's smaller craters are simply holes in the soil (regolith) made by impacting meteoroids that were too small to penetrate as far as the bedrock. Plenty of these types of craterlet are visible on Apollo photographs taken on the Moon's surface. Samples taken from these pits found that they are lined with tiny beads of glass formed from regolith material melted by the impact, and some fragments of the disintegrated meteorite may lie buried beneath the crater's floor and scattered about the surrounding regolith. Most craters larger than 30 metres were produced by impactors that

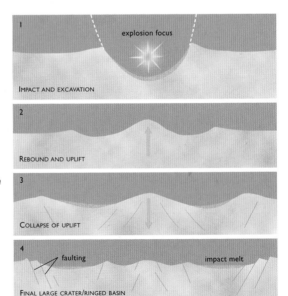

▶ The formation of a large impact crater or ringed basin.

1 The impacting body vaporizes inside the lunar crust and explodes, excavating a large amount of melted material. The crust beneath and surrounding the explosion focus is compressed. A sheet of vaporized material is thrown out around the crater.
2 The crust rebounds and thrusts upwards.
3 The central uplift and rim collapse.
4 The feature settles, impact melt levels out, and faulting occurs.

1 — explosion focus — IMPACT AND EXCAVATION
2 — REBOUND AND UPLIFT
3 — COLLAPSE OF UPLIFT
4 — faulting — impact melt — FINAL LARGE CRATER/RINGED BASIN

managed to dig down sufficiently through the regolith to strike the solid lunar bedrock. Such craters are generally shallower in relation to their diameter. They display raised rims and are surrounded by sheets of ejected material several times the crater's diameter. Apollo astronauts sampled the environments around several small craters that had bored through the regolith to shatter the lunar crust, and particles of bedrock thrown out by the impact were readily located.

A small lunar impact will distribute ejecta in a familiar pattern. Material that once lay nearest the surface will be deposited at the greatest distance from the impact site. Underlying fragments of bedrock will be scattered closer to the crater, and much of it will be heaped up to form a substantial blocky rim. Most crater cavities have been infilled to some extent by ejecta that have fallen back in, along with subsequent processes of soil creep (the gradual migration of loose surface material down a slope) and erosion. This means that much of the pre-cratering layers of regolith and lunar rock are obscured by piles of debris.

The transition from simple mechanical crater excavation to explosive crater formation occurs when large, high-velocity asteroids impact the Moon with enough energy to blast out craters larger than 2 km across, a size at the very limit of most amateur telescopic resolution. The ejecta systems of such impact craters incorporate secondary craters that were gouged out by chunks of material thrown out by the original impact. As much as 75% of the ejecta blanket arises from the churning effects in outlying areas of regolith caused by huge piles of falling debris and secondary impacts. Larger craters may be surrounded by radial furrows, many of which take the form of crater chains or elongated craters, evident around features such as the 93 km diameter Copernicus. Radial and concentric ridges flank many of the Moon's larger impact features, and the inner walls are often around twice as steep as the outer flanks.

Large lunar craters can have a depth/diameter ratio of as much as 1:30, a fact that may surprise anyone who has observed craters near the Moon's terminator and imagined that they were incredibly deep, steep-sided hollows. To anyone standing on the surface of the Moon, the summit of a 3000-metre mountain would disappear below a flat horizon at a distance of just 102 km. In the case of the largest craters, it would be possible to stand on the rim and be unable to see across to the opposite side because the curvature of the Moon's surface takes it far below the visible horizon.

Central hills or mountain peaks are found on the floors of many large impact craters. They are produced by the rebound that occurs after a large impact. When the vaporized impactor and superheated rock that forms around it explode deep inside the Moon's crust, the resilience of the bedrock makes the outward force act more easily to the sides and upwards from the focus of the explosion, creating an expanding torus-shaped region of material. This will tend to carve out a doughnut-shaped crater base from which there is an upward thrust immediately after impact, the underlying rocks experiencing a degree of elastic rebound. Very large impact features show multiple rings rather than central uplifts, and one of the largest craters with a central massif is the farside crater Tsiolkovsky, some 185 km in diameter. Once a large crater has been formed, the lunar crust begins to respond to a changing set of stresses. Material may slump from the rim on to the inner slopes and the floor. Multiple terraces often result from these large-scale landslips. Through a telescope, the extremely complicated interior walls of some large craters can be observed, the interior of the crater Copernicus being a prime example.

Copernicus and Kepler

As the Moon's sunrise terminator sweeps across its western hemisphere, when the Moon is about 8 days old, the broad eastern flanks of the spectacular 93 km diameter crater Copernicus are illuminated by the rising Sun. Copernicus appears as a notch on the terminator, visible with the naked eye by keen-sighted observers. Through a telescope, the intricacies of Copernicus' massive inner and outer walls

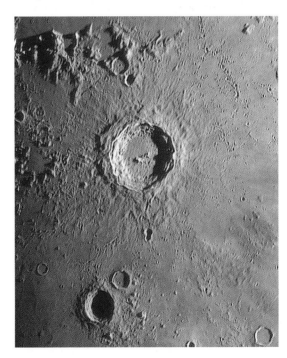

▶ Copernicus, in an image obtained on 4 March 2001 by Mike Brown, using a 370 mm Newtonian reflector and a Starlight Xpress HX516 CCD camera.

are nothing short of breathtaking, with concentric ridges lying close to the crater's rim that give way to radial hummocks farther away. About 50 km south of Copernicus' rim lies the 15 km long keyhole-shaped crater Fauth, named after the German selenographer Philip Fauth, who made accurate and detailed visual maps of the Moon a century ago. Fauth is probably a secondary crater caused by the impact of a large mass of material thrown out when Copernicus was formed. As the terminator edges westwards, Copernicus is revealed in all its majesty. Dotted around can be seen hundreds of smaller secondary impact craters, many of which lie along distinct radial lines, are elongated and are in some cases interconnected. A mighty system of ridges and furrows surrounds Copernicus and radiates outwards for more than 100 km, leading on to the spectacular system of rays that track unimpeded across the surface to distances of several hundred kilometres. At full Moon, the fully illuminated Copernicus and its impressive ray system can grasp the observer's attention for hours on end.

Copernicus is the star of perhaps the first truly spectacular image returned from space. An oblique view of the crater from the cameras of Lunar Orbiter 2 in November 1966, taken from a height of 130 km above the Moon's surface some distance south of Copernicus, was hailed by some as the 'picture of the century'. The image captures the true dimensions of Copernicus – not an immensely deep bowl-shaped cavity, but a vast circular plain sunk 3750 metres below a rim that itself rises 400 metres above the surrounding plain. Its rounded cluster of central mountains rises to heights of more than 1000 metres above the crater's floor. Lunar Orbiter's close-up view shows tremendous detail on the floor and in the central mountains, and along Copernicus' complex northern wall can be seen ridges and gorges, formed by material that has slumped down to the base of the walls. Giant slabs of rock have slipped down from the upper slopes of the central mountains. In 1967, the 28-member geology working group of the Santa Cruz conference suggested that a future Apollo mission might land near Copernicus' central mountains. Sadly, this mission never came to fruition, though it would surely have been Apollo's most visually spectacular landing location.

Copernicus was blasted out of the lunar crust around 900 million years ago by an asteroid measuring up to 10 km across. The 32 km diameter crater Kepler, 500 km to the west of Copernicus, was formed at around the same time. Copernicus and Kepler are separated by a hummocky plain across which the ray systems of both craters intermingle. Although Kepler appears to have a Copernican ray system in miniature, close telescopic scrutiny shows it to be by no means a mini-Copernicus since it lacks a big central mountain massif and complex terraced walls. Kepler's rim is sharp, the walls rising

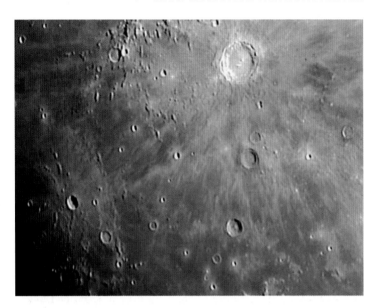

▲ *Impact craters Copernicus (top right) and Kepler (lower left) in an image obtained on 23 April 2002 by Brian Jeffrey, using a 102 mm refractor and a Phillips ToUcam Pro webcam. The rays are clearly visible in this image.*

some 2300 metres above the floor. The outer walls are concentric and layered to a certain extent, but there is only a trace of internal terracing. Kepler is very slightly polygonal in form. Through a small telescope the features on Kepler's floor are not prominent, but a high-power view through a 150 mm reflector will show numerous low, rounded hills. The Clementine probe returned images showing intricate detail within and around Kepler. Along the rim there appear to be narrow gullies, just like those created by water erosion at the rim of the Barringer Crater in Arizona. However, the cause of such lunar features was not water, but rather the action of landslides, slumping and moonquakes.

Viewed at low power through the telescope, Kepler's rays appear pretty uniform in tone, whereas Copernicus' rays are patchier. Both of these old ray systems are overlain with the fresher ejecta of Aristarchus to the north-west (deposited perhaps as recently as 300 million years ago). Kepler's rays occupy a total area of around 50,000 sq km, and some spindly ray fingers can be traced to distances of 300 km or more. There is little telescopic evidence for secondary cratering by ejecta from Kepler, as there is with Copernicus, though the many bright craterlets in the vicinity may have been produced in this manner. Under a low angle of illumination, numerous low ridges and hills can be observed in Kepler's vicinity. Some of these features are aligned radially to Kepler, giving the appearance of being huge mounds of deposited ejecta, or hills that were severely scoured and sculpted by the blast of the Kepler impact. However, these relief features actually pre-date Kepler by billions of years, and are instead remnants of early lunar crust that escaped being covered by the basalt lava flows that formed Oceanus Procellarum, the largest of the lunar maria, which erupted more than 3 billion years ago. Under a high angle of illumination there is clear evidence of a dark collar around Kepler's rim. Once thought to be a ballistic shadow zone (an area which escaped being covered by ejecta), the dark collar is actually made up of a smooth glassy impact melt which was deposited immediately after impact. A number of other relatively young impact craters display such features, most notably the crater Tycho in the southern highlands.

Tycho

To the unaided eye, the southern hemisphere of the Moon appears to be the brightest part of the lunar surface. Tycho can be discerned as a brilliant spot just below the mouth of the 'Man-in-the-Moon' (Mare Nubium). Full of impact craters, the Moon's southern uplands are spectacular to view through binoculars or telescopes. Some of the craters, like Sasserides (90 km across, just north of Tycho), are highly eroded and very ancient. Tycho, 85 km in diameter, is one of the youngest large lunar craters, formed by the impact of a small asteroid, perhaps around 100 million years ago. Its floor lies nearly 5 km beneath the clean-cut rim, and its bright central peaks rise to

◀ Tycho (above centre) and Clavius (below centre) in the lunar southern uplands. The image was obtained on 9 September 2000 by Peter Grego, using a 150 mm refractor and Ricoh RDC-5000 digicam.

1600 metres above the floor. Tycho still looks remarkably fresh. Its impressive system of rays radiates for hundreds of kilometres. Though they may appear to be thin and possess an almost translucent quality, in places the rays are many tens of metres deep.

Larger craters (often referred to as 'walled plains')

According to one of the many criteria outlined by the 19th-century English amateur selenographer Edmund Neison, walled plains are features over 70 km across that have low walls, moderately sunken floors and often lack central mountain structures. However, walled plains are not a distinct species in themselves – they are simply large impact craters whose floors have been filled with lava to a greater or lesser extent, obliterating many of the traces of the original floor and central elevations. It was a popular belief that many of the walled plains lay along lineaments, and supporters of this view claimed that it indicated that the walled plains were of internal origin, rather than being the result of random bombardment. In truth, the so-called lineaments were the combined result of fanciful speculation about lunar tectonic activity and seeking patterns in a random landscape. Certain lunar lineaments do exist, but they are not volcanic: they are largely the product of basin formation and the resultant stresses imparted to the crust around the site.

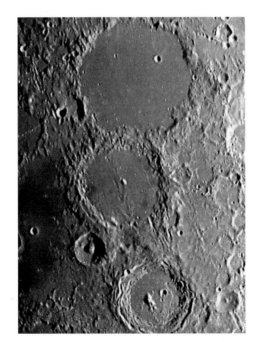

▶ The magnificent walled plains (from top) Ptolemaeus, Alphonsus and Arzachel. The image was obtained on 24 April 1999 by Mike Brown, using a 370 mm Newtonian reflector and a Starlight Xpress HX516 CCD camera.

Many superb walled plains are visible through small telescopes. Several of them, including the 225 km diameter Clavius, are so large that some eagle-eyed observers claim to be able to pick them out without optical aid when they lie near the terminator. Plato (100 km), a beautiful walled plain sunk into Montes Alpes, the lunar Alps, has a dark flat floor upon which several craterlets are visible through telescopes with apertures greater than 150 mm. Archimedes (83 km) lies in Mare Imbrium about 500 km due south of Plato. Endymion (125 km) is a lovely dark-floored walled plain near the Moon's north-eastern limb. Near the north-western limb lies J. Herschel (156 km), an ancient and considerably eroded walled plain. Grimaldi (222 km) is the most prominent walled plain visible at full Moon, situated near the western limb and apparently visible to the naked eye by those who possess excellent vision. Albategnius (136 km), Ptolemaeus (153 km) and Alphonsus (118 km) are situated near the centre of the Moon's disk and all three (often referred to as the 'Ptolemaeus chain') appear quite spectacular under low illumination. Farther south are to be found the walled plains Purbach (118 km), the misshapen Regiomontanus (126 × 110 km), the highly eroded plain of Deslandres (234 km), and Walter (140 km) with its fascinating cluster of floor craters. Schickard (227 km), near the south-west limb, is famous for its multi-toned floor, the outcome of several phases of volcanic infilling. The vast ancient plain of Bailly (303 km) is the largest walled plain on the Moon, some 300 km south of Schickard and close to the lunar limb.

Impact basins and the maria

Large grey plains, clearly visible with the unaided eye, cover more than a third of the Moon's nearside. The larger of these maria, the lunar 'seas', were given romantic names by the Italian astronomer Giovanni Riccioli in his 17th-century map; they are still in use today. The western hemisphere is largely covered by the vast Oceanus Procellarum (Ocean of Storms) and Mare Imbrium (Sea of Rains). Mare Imbrium, the largest circular sea on the Moon, is bordered by an impressive series of high mountain ranges, except in the west where it blends into the plains of Oceanus Procellarum. To the south of Oceanus Procellarum lie two smaller circular seas, Mare Humorum (Sea of Moisture) and Mare Nubium (Sea of Clouds). Oceanus Procellarum extends northwards in the guise of Mare Frigoris (Sea of Cold), and this narrow marial tract arcs north of Mare Imbrium. To the south of Mare Imbrium lie Mare Vaporum (Sea of Vapours) and Sinus Medii (Central Bay). The eastern hemisphere is punctuated by a number of distinct marial basins. Mare Serenitatis (Sea of Serenity) adjoins Mare Imbrium in the north. To its south is Mare Tranquillitatis (Sea of Tranquillity), which branches southwards into Mare Nectaris (Sea of Nectar) and south-east into Mare Fecunditatis (Sea of Fertility). Near the eastern lunar limb is the oval-shaped Mare Crisium (Sea of Crises), the best-preserved of the nearside marial basins.

Although many early lunar observers were keen to promote the Moon as an Earth-like world, few of them actually believed the maria to be vast expanses of water. A low-power telescopic view gives the impression that the seas are flat and featureless (especially at full Moon), but a high-power perusal of the maria in the vicinity of the terminator will clearly reveal that they are full of undulations, ridges, hills and craterlets. The early lunar observers saw no signs of waves lapping at the edges of the seas, nor did they discern white foam frothing along shorelines. For the most part, the seas of the Moon seemed to be unchanging. Sunlight does not glint off the lunar maria, as it would if they were composed of water – they are matt, and obviously composed of solid stuff.

In the late 19th century the American geologist Grove Gilbert championed the theory that the maria are lava flows that occupy large basins excavated by asteroidal impacts. Today we regard the marial basins as huge impact features that have been modified by volcanism, lava flows and subsequent impacts. In effect, there is not much difference between a very large flooded crater and a small marial basin, but most lunar geologists are content to refer to any impact structure with a diameter larger than 300 km as an impact basin, rather than a large crater. There are more than 40 such impact basins on the Moon.

▲ *Sunrise over eastern Mare Imbrium. The image was obtained on 1 April 2001 by Doug Daniels, using a 150 mm refractor.*

Impact basins and mare formation – a history

Much of the asteroidal bombardment suffered by the Moon took place very early in its lifetime, the most violent phase ending about 3.8 billion years ago. Attempts have been made to introduce a geological timescale to distinguish separate eras of lunar history, but establishing a sequence of events and determining the relative ages of the lunar basins has proved less difficult than finding out when these events actually took place. While the sequence of major lunar events is now known, there are not enough positively datable events for lunar geologists to produce a detailed lunar timescale.

THE BIRTH OF THE MOON

Questions about the Moon's origin have been asked for thousands of years. Significant progress began to be made in the latter part of the 20th century, when the Moon's material was sampled directly and its surface imaged and analysed up close. A wealth of data acquired by Apollo and dozens of lunar probes, especially the US orbiters Clementine in 1994 and Lunar Prospector in 1998, has given scientists a great deal of evidence on which to base their theories. Lunar rocks have been dated and their composition identified, telling us about the conditions in which they were created, and even suggesting how the Moon was formed in the first place. No less than 99% of the Moon's surface is older than 2 billion years, and some of the very ancient rocks returned by the Apollo astronauts tell us that the Moon has existed as a single, separate body for around 4.6 billion years.

How did the Earth come to be partnered by such a large satellite, a quarter of the Earth's diameter? What can explain the significant difference in composition between the Earth and the Moon – and the points of similarity? Four major theories have been put forward

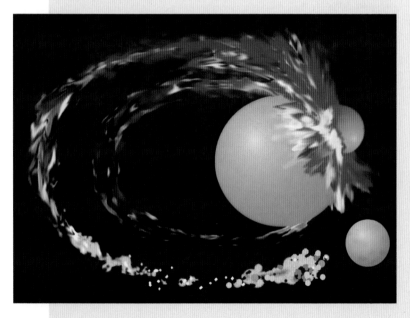

to account for the Moon's origin: fission, capture, co-accretion and impact. According to the *fission* hypothesis, the Moon split away from the Earth as a result of irresistible centrifugal forces. Only in the most extreme (and unlikely) circumstances could this ever have happened – moreover, it does not explain the compositional difference between the Earth and Moon. The *capture* theory envisages that in the distant past the Moon was an independent planet, but was captured by the Earth's gravity. A separate lunar origin provides an explanation for the difference in composition seen between the two worlds, and it also explains why the Moon now orbits the Earth close to the ecliptic. However, if the Moon had once been an independent planet, its momentum upon encountering the Earth would have to have been reduced dramatically for it to have been captured into an orbit around the Earth, and it is hard to see how this could have happened. *Co-accretion* theory has the Moon condensing from a primordial cloud of gas, dust and assorted debris in orbit around the embryonic Earth. This 'sister planet' theory, however, does not account for the big differences in composition between the two objects.

Now widely accepted to be the most likely origin of the Moon is the *giant impact* or '*big splash*' theory. This theory suggests that a Mars-sized planet (around half the size of the Earth) smashed into the young Earth, disintegrating the impactor and the Earth's mantle at the site of impact. A cloud of debris was splashed into near-Earth orbit, and the outer regions of this temporary ring of material coalesced to form the Moon. While most lunar genesis theories find it difficult to explain the dearth of metallic iron thought to exist at the Moon's centre, this one provides a neat answer – the impactor's iron-rich core joined with that of the Earth, and the splashed-out material was composed of the two bodies' relatively iron-sparse mantles. The Apollo missions demonstrated that the Moon has little water or other volatiles (substances which vaporize at relatively low temperatures) in its endogenous rocks. The 'big splash' theory again provides an answer – the impact generated sufficiently high temperatures to boil this volatile material off into space.

◄ *A gigantic Mars-sized planet is thought to have impacted with the embryonic Earth. The core of the impactor joined the body of the Earth, while the mantles of both objects were mixed and splashed out into space. Some of this material returned to the Earth, but a substantial proportion remained in orbit and gradually accreted to form the Moon.*

The *Pre-Nectarian period* spans the time between the Moon's formation, 4.6 billion years ago, to 3.9 billion years ago. Many of the features visible from this period are heavily eroded due to their great antiquity. Halfway through the Pre-Nectarian Period, around 4.3 billion years ago, the Moon's crust was fully formed and soon to be scarred by two of the Moon's most ancient impact basins – the Procellarum basin and the farside South Pole-Aitken basin. These were followed by the impacts that produced many of the farside basins, along with the nearside basins of Tranquillitatis, Fecunditatis and Nubium. The Nectaris basin was formed around 3.9 billion years ago, and marks the beginning of the *Nectarian period*. During this period, the Humboldtianum, Humorum, Crisium and Serenitatis impacts took place. The *Imbrian period* commenced 3.8 billion years ago with the creation of the Imbrium basin, and includes the formation of the Orientale basin.

Large-scale lunar volcanism died out about 3 billion years ago, as the lunar crust thickened and its upper mantle lost its heat and solidified. The *Copernican period*, the most recent period in lunar history, began with the formation of the crater Copernicus about 900 million years ago. No volcanic activity has occurred since then, and the rate of major impacts has dwindled to its current level.

Lava flows spread across the wide expanse of the Imbrium basin, obliterating most of its inner structures, but Orientale's flooding was restricted to its centre and a few outlying areas. This makes the Orientale basin one of the best-preserved major impact sites in the Solar System. The lavas of Mare Tranquillitatis, Mare Crisium and Mare Fecunditatis flowed around 3.5 billion years ago, preceding the infilling of Mare Imbrium and Oceanus Procellarum in the western

▶ The walled plain Plato in the Montes Alpes. Below it is a portion of northern Mare Imbrium, showing Mons Pico (towards bottom centre), Pico Beta (farther south) and (centre left) the Montes Teneriffe. The image was obtained on 21 September 2000 by Mike Brown, using a 370 mm Newtonian reflector and a Starlight Xpress HX516 CCD camera.

hemisphere by several hundred million years. The marial lava is thought to have formed in a hot radioactive decay melt layer – in places 100 km thick – up to 200 km beneath the lunar surface. As the outer shell of the Moon cooled and consolidated, this magma layer gradually retreated deeper into the Moon's mantle. The basaltic lava flows were generally very fluid and fast-moving, giving no time for large quantities to accumulate in localized areas. Samples of mare rock have been melted in the laboratory and found to have the viscosity of motor oil. The lunar maria were built up from layer upon layer of this runny material. More than 30% of the nearside was covered with basaltic lava, compared with less than 3% of the farside. Only a few farside impact basins were flooded with lava because the crust overlying the hot melt layer was significantly thicker, by up to 50%, than that of the nearside, and it experienced fewer really big impacts. Large-scale lunar volcanism had dwindled by around 3 billion years ago, accompanied by a steeply declining impact rate.

Mountain ranges and peaks

Many of the lunar maria are bordered by mountain ranges that rise sharply from the flat mare plains. It is worth noting that despite their jagged, angular appearance through the telescope – especially when they cast long shadows like cathedral spires at low angles of illumination – most of the lunar mountains actually have gentle slopes, on the whole no steeper than 30°.

Mare Imbrium, the largest of the Moon's circular seas, is surrounded for 300° of its circumference (some 3900 km) by the most spectacular and near-continuous series of mountain ranges on the nearside. Clockwise around Imbrium, beginning in the north-west, there are two massive, well-rounded mountain blocks called Mons Gruithuisen Gamma and Delta, both of which have base diameters of 20 km. Gamma is topped by a tiny crater, which indicates that these features are probably large volcanic domes. To the north of these sentinel mountains, cratered hilly terrain around the crater Marian rises from the flat expanses of Oceanus Procellarum to the west and blends into the Montes Jura, a magnificent arc of peaks rising steeply to 2000 metres. The Jura mountain arc forms the northern border of the magnificent bay of Sinus Iridum (Rainbow Bay), a 260 km diameter asteroidal impact basin which has been infilled with lava. The southern half of the original ramparts of the Iridum crater has been obliterated by volcanic activity, but traces of the wall can still be seen, outlined by low wrinkle ridges. An inner Iridum crater ring some 120 km in diameter and marked by faint ridges can also be traced under the right illumination conditions.

Undulating mountains continue to sweep north of Mare Imbrium, broadening into the spectacular Montes Alpes. This range engulfs the large dark-floored Plato, an impressive walled plain 100 km across. In places the peaks of Montes Alpes rise to average heights of 2400 metres. Mons Blanc in the southern lunar Alps rises to an impressive 3600 metres. In Mare Imbrium, south of Plato, several mountain ranges and peaks are entirely surrounded by mare material. Because they are prominent and lie on relatively flat terrain, these features and the shadows cast by them are well placed for telescopic study. Montes Recti (Straight Range) is a linear range of peaks, 90 km long and in places rising to 1800 metres. It is a pity that the Montes Recti are orientated east–west and in line with the Sun's path through the lunar skies; if they ran north–south, their appearance after local sunrise or before sunset would be absolutely breathtaking, casting a broad, jagged band of shadow. Just east of Montes Recti are Montes Teneriffe, a cluster of mountains rising to heights of 2400 metres in places. These peaks mark the western boundary of a barely visible buried crater in Mare Imbrium, unofficially called 'Ancient Newton'. East of Montes Teneriffe is the prominent Mons Pico, which rises to 2400 metres above its surroundings, and about 35 km to its south lies a smaller, bone-shaped mountain called Pico Beta. Montes Teneriffe, Mons Pico and Pico Beta, interwoven with some of the marial wrinkle ridges, make a wonderful sight when illuminated by an early morning or late evening Sun.

In the west of Mare Imbrium, the 2250-metre-high Mons Piton stands proud and solitary above its grey surroundings. All these moun-

◀ The Montes Apenninus and a portion of eastern Mare Imbrium, showing the Palus Putredinis. The walled plain Archimedes is at top left. This image was obtained on 13 February 2000 by Mike Brown, using a 370 mm Newtonian reflector and a Starlight Xpress HX516 CCD camera.

tain features are thought to predate the lava infill, and are the surviving traces of the Imbrium basin's inner ring. About 80 km east of Piton, at the end of Montes Alpes, lies the crater Cassini with its unusual extended outer flange and its dark flooded floor punctuated with two large craters. West of Cassini are Montes Caucasus; a substantial range 550 km long, the Caucasus mountains mark the boundary between north-western Mare Serenitatis and eastern Mare Imbrium. In the western part of Mare Imbrium, some 50 km north of the walled plain Archimedes, the Montes Spitzbergen mountain group makes a fascinating object for study. Roughly 60 km long, with peaks up to 1500 metres high, this neat cluster of mountains was named by the English selenographer Mary Blagg for their resemblance to the terrestrial Spitzbergen islands in the Arctic Ocean. The region south of the crater Archimedes is populated by Montes Archimedes, an area of chaotic mountainous terrain covering around 45,000 sq km; those with very good eyesight can actually discern the Montes Archimedes region with the naked eye as a light patch.

Mare Imbrium's south-eastern border is formed by Montes Apenninus, the lunar Apennines. Some individual peaks within this vast range rise above 5000 metres; the enormous 70 km long mountain block of Mons Huygens reaches 5400 metres high. Much of the Apennine range is striated in a pattern radial to Mare Imbrium – a clear imprint of the forces that formed the impact basin and the subsequent crustal faulting. The eastern end of the lunar Apennines is marked by the 60 km impact crater Eratosthenes, after which there is a gap in the mountain chain. It is picked up 90 km farther on by the Montes Carpatus (Carpathians), which mark Mare Imbrium's southern border. The Carpathians are over 400 km long and are overlain by thick deposits of bright material ejected by the Copernicus impactor some 900 million years ago.

Though Mare Imbrium and its borders provide the lunar observer with a rich hunting ground for ranges and isolated groups and peaks, there are many more magnificent mountains to be found in other regions of the Moon's nearside. The southern sector of Oceanus Procellarum is a wonderland of peaks, mountain ranges, domes, wrinkle ridges and buried craters. It is dominated by Montes Riphaeus, an impressive hook-shaped mountain massif 150 km in length. Sprawling across the south-eastern part of Mare Humorum is the Promontorium Kelvin, a multi-faceted pyramid with a base area of around 1000 sq km, rising to a height of 3000 metres. In Mare Nubium, south of the famous fault known as Rupes Recta (Straight Wall), is a superb little cluster of peaks known as the 'Stag's Horn Mountains'. Although they are rather small in comparison with the features discussed earlier, their unusual shape and location near one of the Moon's most notable fault features make them an unforgettable sight.

Wrinkles in the lunar seascape

Here and there the maria are crossed by low ridges known as wrinkle ridges. These ridges are just tens of metres high and so are visible only when illuminated at a very oblique angle, just after local sunrise or before sunset. Their mode of formation is still something of a mystery. The paths of most wrinkle ridges tend towards being concentric or radial to the maria. It has been suggested that they are banks of viscous lava which erupted through deep crustal fissures. Most wrinkle ridges are thought to be compression features; they probably formed when the mare lavas subsided, reducing the marial surface area and crumpling mare material in a gradual manner. There is strong evidence for marial contraction from spaceprobe photographs of mountain borders

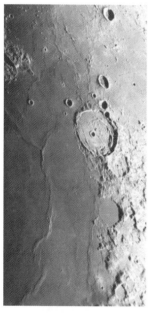

▲ *The eastern border of Mare Serenitatis, showing Posidonius and the giant wrinkle ridge Dorsa Smirnov. The image was obtained on 11 March 2000 by Mike Brown, using a 370 mm Newtonian reflector and a Starlight Xpress HX516 CCD camera.*

– 'high tide' marks, which must have been etched during early lava flooding, are clearly visible. Broader lava fronts can also be seen in places on the maria. Some of the wrinkle ridges seem to follow the pre-existing outlines of flooded inner basin rings, and are therefore a reflection of the underlying topography.

Most of the Moon's wrinkle ridges are named in honour of prominent naturalists and geologists. Perhaps the best-known are the Dorsa Smirnov (once called the Serpentine Ridge), an easily observable system of ridges which runs parallel to the eastern rim of Mare Serenitatis for around 130 km. Dorsa Lister run around the southern part of this mare, through the crater Bessel, to join with the Dorsum Azara. Farther west are the Dorsa Buckland, von Cotta, Gast and Owen. There are several prominent (currently unnamed) wrinkle ridges in the north of Serenitatis. Crisium, Humorum, Nubium, Fecunditatis and Tranquillitatis all have extensive systems of these ridges. Wrinkle ridges make up an inconspicuous – though unique and fascinating – feature in south-western Mare Tranquillitatis called Lamont. This 75 km diameter feature is composed of a series of low wrinkle ridges, giving it the appearance of a bullet hole in a sheet of toughened glass. Lamont seems to defy general classification. It cannot really be called a crater, although the main feature is roughly circular in outline. A system of ridges spreads outwards across the mare floor to a distance of 100 km. Could Lamont be a relic of Mare Tranquillitatis' contraction and slumping, or perhaps an amazing example of a volcanic crater that never reached maturity? Or is it an ancient impact crater surrounded by a hefty ejecta collar that was completely buried under young lava flows? Just south of Lambert in Mare Imbrium there is another ghost feature, Lambert R, which is is likely to be similar in nature to Lamont.

Domes

Low rounded hills known as domes occur singly or in small clusters in parts of the maria. Like wrinkle ridges, domes rise very gently from the landscape and can only be seen when near the terminator, illuminated by a low Sun. Many domes are capped by tiny summit craterlets, the majority of which are beyond the reach of small telescopes. It is thought that most domes are volcanic features built up by successive eruptions of lava. (The lava may have had a viscosity equivalent to that which formed, and is still forming, the Hawaiian shield volcanoes on Earth.) The craterlets that crown the domes are likely to be the remnants of long-cooled volcanic vents. It is possible, though, that some domes are uplift features like the Adirondack Mountains in New York State, where deep intrusions of magma have forced the overlying layers of rock to arch upwards. A lunar feature similar in size to the Adirondacks is the impressive Mons Rümker in northern Oceanus Procellarum, a distinctly lumpy sprawling plateau 70 km in diameter rising above a relatively flat plain. The feature is best seen when it has just emerged into the lunar morning, and it will repay close telescopic scrutiny.

Some domes are clustered together, for example the dome fields west of Copernicus, the domes north of Hortensius and those lying west of Marius. On Earth, superficially similar features are the drumlins

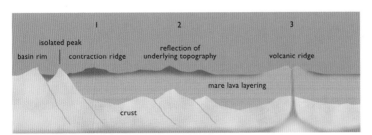

▲ A cross-section through part of a large flooded ringed basin, showing three ways in which wrinkle ridges (dorsa) might form.
1 The mare slumps and a contraction feature forms.

2 The mare slumps and surface wrinkles are a reflection of the underlying topography.
3 A lava ridge forms as a result of volcanic activity along a fault line.

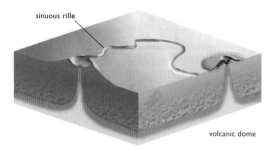

◀ *A cross-section through part of a mare, showing the formation of a sinuous rille and a volcanic dome. The rille is formed by lava flow from a volcanic vent. The dome is built up from layers of erupted viscous lava and ash.*

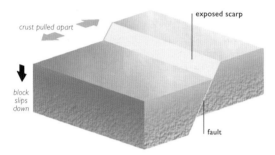

◀ *Tension in the lunar crust gives rise to faulting and the formation of a fault scarp.*

– so-called 'basket of egg' topography – caused by glaciation. Clusters of lunar domes formed in areas of crustal weakness, in places where the crust had been fractured by major impact or extensively faulted before being buried by the marial lava flows. Surprisingly, only a small number of lunar crater floors contain appreciable domes. The best example is the floor of 60 km Capuanus, which boasts a group of several low domes. The floor of the crater Mersenius (84 km) is itself a dome of sorts: it is noticeably convex even when observed through small instruments, and a small dome lies at its northern end.

A faulted Moon

By around 3 billion years ago, large-scale lunar volcanism had ceased. The molten lava outpourings that formed the lunar maria dwindled, and the violence of the Moon's early history gave way to a much quieter phase, involving occasional impact events and global crustal adjustment. The Moon began to settle down after a difficult adolescence, and faults appeared in areas where tension and compression overcame the natural strength of the lunar crust. Most of the major lunar faults visible today are believed to have been produced between 3 billion and 1 billion years ago, when the lunar crust had gradually cooled and begun to settle down to achieve a state of equilibrium.

There are dozens of lunar faults that are easy to observe through a small telescope, and a good 150 mm reflector is capable of revealing literally hundreds of them. 'Normal' faults are the simplest type, result-ing from tension forces that have pulled the crust apart. Where the crust has cracked, the force of gravity has produced a horizontal displacement between the two separated blocks; the upper block's exposed edge is called a fault-scarp, or rupes. By far the best example of a normal fault on the Moon is Rupes Recta, the Straight Wall, in south-eastern Mare Nubium. This fault-scarp has a rather gentle gradient of 7° and runs north–south for 126 km from the Stag's Horn Mountains to the craterlet Birt D in a very slight curve. Rupes Recta is best seen just after first quarter phase, when it throws a prominent broad shadow westwards on to the relatively flat mare floor, but it is virtually invisible at a high angle of illumination, when no shadows are cast. Around last quarter, however, the scarp face shows up clearly as a bright narrow line. Parallel to Rupes Recta and 35 km to its west is a narrow cleft known as Rima Birt, which connects the craterlets Birt E and Birt F.

In the west of Mare Fecunditatis, the shadow cast by an unnamed normal fault 60 km long with an eastward-facing escarpment is easy to find three days after full Moon, since the searchlight-like rays emanating from the crater Messier A point directly towards it. In the neighbouring Mare Tranquillitatis, the 180 km long south-west-facing scarp of Rupes Cauchy is a good target for small telescopes when the Moon is four days old. In places the feature 'lapses' into a rille. This area has obviously undergone substantial crustal adjustment, since the Rima Cauchy runs parallel to the Rupes Cauchy at an average distance of 50 km north of it.

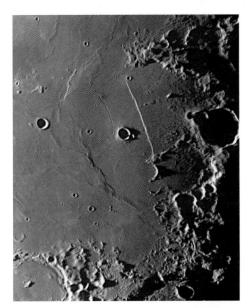

◀ *Rupes Recta in Mare Nubium. The image was obtained on 21 September 2000 by Mike Brown, using a 370 mm Newtonian reflector and a Starlight Xpress HX516 CCD camera.*

Some of the Moon's faults do not have such clean textbook forms. The Rupes Altai is a gigantic scarp of staggering proportions; it traces a scalloped arc for 480 km parallel to the south-western border of Mare Nectaris, from near Catharina to Piccolomini. The Altai scarp actually represents the remnants of the 860 km diameter primary ring of the Nectaris basin. On the opposite side of the Nectaris basin, crustal tension has given rise to a series of narrow rilles – the Rimae Gutenberg, visible only when illuminated by a low Sun.

Rilles

It became evident to the earliest lunar observers that there is much more to the Moon than seas, mountains and craters. With each advance in telescope optics there came better resolving power, making it increasingly apparent that the lunar surface is exceedingly intricate. With their state-of-the-art telescopes, 18th-century observers began to search for an elusive class of lunar feature, known examples of which had previously been few and far between – narrow valleys, some of which were straight, others curved, and some sinuous. These valleys were studied in detail in the late 18th century by the German amateur astronomer Johann Schröter, who called them rilles (*rille* being German for 'groove'), a word still in extensive use by today's lunar and planetary geologists.

Straight rilles are found in both flat and mountainous lunar terrain. They are typically more than 5 km wide and can stretch for hundreds of kilometres, cutting across maria and upland alike. Their origin appears to be very deep-seated, as they are found where crustal tension has caused close parallel faults to form and the land in between to subside. Any pre-existing craters and mountains in the path of the faulting are deformed by it. Photographs taken from lunar orbit show many striking examples of rilles that cut cleanly through crater walls, causing a vertical displacement of the crust along its length. A similar geological process formed the East African rift valley system (6400 km long, average width 50 km), and though

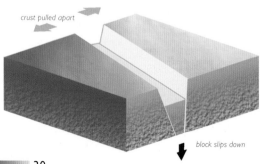

crust pulled apart

block slips down

◀ Tension in the lunar crust gives rise to parallel faulting. The central block sinks below the mean level of the lunar surface, and a graben rille is formed.

◄ *The Triesnecker (centre) and Hyginus (top right) rilles in the Sinus Medii. The image was obtained on 19 October 2000 by Mike Brown, using a 370 mm Newtonian reflector and a Starlight Xpress HX516 CCD camera.*

there is nothing on the Moon to compare with it in terms of sheer scale, there are plenty of well-defined rilles that the lunar observer can explore.

The best example of a lunar rift valley – though it is so wide that it is not classed as a rille – is Vallis Alpes (Alpine Valley), a unique feature that slices cleanly through 130 km of Montes Alpes. In places, Vallis Alpes is 20 km in breadth, and on average its steep walls rise some 2000 metres above its floor. A small rille winds its way down the centre of the valley floor. Vallis Alpes is so neat and tidy that it looks as though it has been carved out of the Moon's crust by some gigantic chisel. I was once told by a respected amateur astronomer that this was actually the case – the 'giant chisel' in his rather dubious scenario being an asteroid which ploughed into the Moon at an incredibly oblique angle! What really happened was that soon after the Imbrium basin was blasted out by an asteroidal impact, the surrounding crust began to settle and parts of the crust succumbed to tension. Two parallel faults appeared right across the Alpine mountain range, the primary rim of the Imbrium basin. The area bounded by the faults subsequently sank down beneath the level of the mountaintops. Geologists call this kind of valley a graben (German for 'ditch'). The floor of Vallis Alpes was later flooded with lavas that flowed from the maria. Finally, a smaller rille appeared down the middle of the valley's floor, carved out by the action of lava flows from Mare Imbrium. Faulting occurred as the terrain in and around the Imbrium basin began to settle, and forces of tension and compression deformed the lunar crust.

There are many other examples of fault valleys cutting neatly through the mountain borders of maria. East of the crater Maupertuis are the Maupertuis rilles (Rimae Maupertuis), visible through larger instruments under good seeing. Amid the Alps east of Plato are numerous rilles, the Rimae Plato. Many rilles are visible within the seas themselves, running concentric to them just offshore. Mare Imbrium boasts the Rimae Bradley, Fresnel and Hadley near its south-eastern shore.

A beautiful trio of rille systems lies near the centre of the Moon. Of these, Rima Ariadaeus is the simplest, a 220 km long rift which runs east–west, cutting cleanly through craters and hills. In places the rille is 5 km wide. The western end of Rima Ariadaeus blends into the flattish moonscape and is linked by a diagonal rille to the eastern section of Rima Hyginus. This unique lunar feature, 220 km long and composed of a series of linked craters and pits, is probably a deep-seated fault line that has attracted volcanism and vent crater formation. South of Hyginus are Rimae Triesnecker, an interlacing network of winding clefts which makes a splendid subject for telescopic study just before first quarter, when it is close to the lunar terminator.

Many lunar rilles are concentric to the rims of certain maria. As with Rupes Altai, their presence is not coincidental, for their origin must be very deep-seated. Examples include the system around the crater Goclenius in Mare Fecunditatis, the Rimae Hypatia in Mare Tranquillitatis and Rimae Plinius in Mare Serenitatis.

Crater clefts

Plenty of rilles and clefts are confined solely to the interior of craters. A 150 mm reflector will show that the floors of Gassendi and Posidonius are both scored with networks of fault valleys – evidence that there has been considerable slumping and cracking of the floor material. The crater Petavius is host to several rilles which protrude from the central mountain massif; by far the most impressive of these is the large straight rille that runs 50 km from the central peaks to the inner south-western wall. Near the western lunar limb, the 106 km diameter crater Hevelius contains a fascinating grid of linear rilles. South-west of Hevelius, closer to the lunar limb, the 146 km diameter crater Riccioli shows a similar pattern of linear rilles. Both craters require detailed telescopic scrutiny at a favourable libration (see Chapter 2) and suitable angle of illumination to fully appreciate the complexity of their floors.

The western sector of the weird-looking Lacus Mortis (Lake of Death) is crossed by numerous fault valleys, the most prominent of which is the 100 km long Rima I Bürg, which cuts across the western part of its floor. The fault is intersected by a near-linear normal fault which runs perpendicular to it across to the southern wall of the lake. North of Rima I Bürg are some interesting features, including a 20 km long crater chain which requires a large telescope to resolve adequately.

Winding rilles

One of the loveliest examples of a sinuous rille is Rima Hadley, which runs for about 100 km across the flat floor of Palus Putredinis (Marsh of Decay) near the base of the Apennine mountains. Of all lunar rilles, it is easily the best known – in the popular mind as well as scientifically – because in July 1971 it was visited by Apollo 15 astronauts Dave Scott and James Irwin. Rima Hadley is, on average, some 350 metres deep and over a kilometre wide. From its rim, the astronauts surveyed its interior slopes and noted substantial piles of debris at the base of the rille, comprising soil and large angular boulders up to 30 metres across. Ancient bedrock was exposed along the walls of the rille, and the samples of lunar rock obtained from this location provided a rare insight into the very early history of our satellite.

Rima Hadley, along with many other sinuous rilles, is an example of a feature that was produced by a fast-running river of molten lava. There is some debate whether the lavas were exposed, or whether they flowed within lava tunnels whose roofs subsequently collapsed. Most sinuous rilles appear to begin at a volcanic vent (sometimes atop a dome), follow a downhill course and dwindle in size. They are thought to represent some of the last features produced by lunar volcanism.

Vallis Schröteri (Schröter's Valley) is the finest example of a sinuous lunar rille. It begins at a modest-sized crater pit, widens into the formation known as the Cobra's Head, then narrows again and snakes for 200 km across the mare. Like Vallis Alpes, Vallis Schröteri possesses a narrow, meandering medial rille which may have been cut out of the lunar surface by the action of rapidly flowing lava. This feature was only adequately recorded by the US Lunar Orbiter mapping probes, but a small section of it, where it resides in the Cobra's Head, can just about be glimpsed in a 200 mm telescope. Also in this area is an interesting cluster of rilles known as Rimae Aristarchus, around 70 km to the north-east of Vallis Schröteri. A mass of distinctly serpentine rilles, some sinuous and some more linear, Rimae Aristarchus can be seen in a 150 mm reflector in excellent seeing conditions. Many of these narrow clefts appear to be volcanic channels originating in small craterlets.

From this brief survey of lunar geology, it should be clear that the Moon is one of the Solar System's most geologically diverse bodies. With its giant multiringed basins circled by mountains, thousands of large craters, lava plains and lava-cut valleys, volcanic domes and crustal faults, this amazing geological variety is readily visible through small amateur telescopes. Even though volcanic activity on the Moon died out billions of years ago, and no major impacts have occurred for millions of years, many of the Moon's features are so well preserved that they look to us as if they were formed yesterday.

THE MOON IN SPACE

At 3476 km across, the Moon is about one-quarter the diameter of the Earth. It orbits at an average distance of 384,401 km, about 60 times the Earth's radius, and completes an orbit around the Earth in just under four weeks, going through a complete cycle of phases from new Moon, through first quarter, half Moon and last quarter, to the next new Moon. The shape of the Moon's orbit is slightly elliptical, and drawn to scale it looks almost circular. Every month, the Moon approaches the Earth to within 360,000 km and recedes to more than 400,000 km. The point on the orbit nearest the Earth is called perigee (from Latin *peri-* = 'near', *ge* = 'Earth') and the farthest point is called apogee (Latin *apo-* = 'far').

Like all other celestial motions, the Moon's orbit follows Johannes Kepler's laws of planetary motion, which describe how planets move around the Sun (and satellites around planets) in elliptical orbits. Isaac Newton's theory of universal gravitation explains exactly what keeps one celestial body in orbit around another. Every particle of matter attracts every other particle with a force that depends on their masses and on the distance between them. The more massive a planet, the stronger its gravitational pull; and the attraction experienced by a falling apple or an orbiting satellite falls away with increasing distance from the planet. At an average distance of 384,401 km, the Moon experiences a considerable gravitational pull towards the Earth, which is more than 80 times as massive as its satellite. In turn, the

LUNAR DATA	
Diameter	3475 km
Circumference at equator	10,920 km
Surface area	37,960,000 sq km
Apparent diameter	31' 5" (average), 33' 29" (at perigee), 29' 23" (at apogee)
Distance from Earth	384,401 km (average), 356,400 km (at perigee), 406,700 km (at apogee)
Light distance to Moon	1.3 seconds
Eccentricity of Moon's orbit	0.0549
Magnitude of full Moon	−12.55
Inclination of Moon's orbit to ecliptic	5° 8' 43"
Sidereal month	27d 7h 43m
Synodic month	29d 12h 44m
Average orbital velocity around Earth	3681 km/h
Average angular velocity in sky	33' per hour
Maximum surface temperature at night	−155°C
Maximum surface temperature at midday	105°C

▶ *The changing appearance of the Moon's illuminated disk results from its motion around the Earth. At new (right), the Moon lies in line with the Sun, and the hemisphere presented to Earth is completely dark. At full, the Moon lies opposite the Sun in Earth's sky and is completely illuminated. Between these two extremes, varying degrees of illumination are seen.*

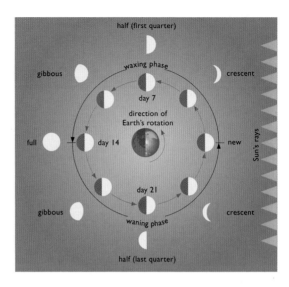

Moon exerts a gravitational pull upon the Earth of equal magnitude (following Newton's 3rd law which states that 'To every action there is an equal and opposite reaction'). While the magnitudes of the forces that the Earth and Moon exert on each other are the same, the effects are not, since Newton's 2nd law states that 'The acceleration of an object is proportional to the net force exerted on it and inversely proportional to its mass.'

The Moon's orbit

Both the Earth and Moon revolve about their common centre of gravity, a point called the barycentre. If the Earth and Moon were equal in all respects then the barycentre would be positioned exactly halfway between the two bodies. But the Moon has little more than 1% of the Earth's mass, which offsets the barycentre considerably towards us – so much so that it is actually located within the Earth, around 4700 km from the centre.

At apogee the Moon can be as distant as 406,700 km, while at perigee it can approach to 356,400 km. This difference of 50,300 km means that the Moon's apparent perigee diameter is 33′ 29″, compared with 29′ 23″ at apogee. In other words, the perigee Moon is larger in apparent diameter than the apogee Moon by 14%. Such a variation can easily be detected with the naked eye using a carefully constructed cross-staff, an instrument commonly used by astronomers in the pre-telescopic era. A lunar cross-staff should consist of a long, sturdy piece of wood with a small sight mounted at one end (a 'squint hole') and a slider at the other end. The slider has two pins

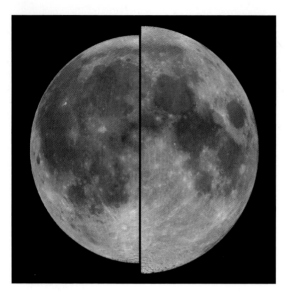

◄ A comparison between the Moon at apogee (left) and perigee (right). The perigee Moon is about 4' wider than the apogee Moon, and presents an area nearly 30% larger.

positioned close together so that the whole lunar disk just fits in between them when sighted though the squint hole. Variation in the apparent diameter of the Moon means that the slider's position requires adjusting slightly each night, and its position can be noted against a scale drawn on to the staff.

Time and tide

The gravitational attraction between the planets and their satellites produces tidal forces in them. Most of the effects are subtle and immeasurable, but some are truly spectacular. Io, the innermost of Jupiter's Galilean satellites, is being gravitationally pumped by the attraction of mighty Jupiter and the other Galileans, raising 100-metre amplitude tides in its solid crust and causing intense frictional heating and permanent volcanism. Although the tides raised on the Earth by the Moon (and to a lesser extent by the Sun) have less dramatic consequences, they are immediately observable, they affect life on Earth in many ways, and they have profound cumulative effects over long timescales.

The tidal force exerted by the Moon has a greater effect on the oceans on the hemisphere of the Earth facing it, and a lesser effect on the oceans on the side of the Earth opposite the Moon. (The Moon's gravity also induces a small tidal effect on the solid body of the Earth.) In effect, two oceanic bulges are produced, one large one facing the Moon and another smaller bulge on the opposite side of the Earth. Since the Earth rotates once on its axis every 24 hours, and the Moon takes about four weeks to revolve around the Earth, friction is generated between the fast-spinning body of the Earth and the oceanic tides. Acting as a brake, this friction gradually slows the Earth's rate of spin (lengthening each terrestrial day by two ten-millionths of a second) and pushes the tidal bulges forwards, in the direction of the Earth's rotation. A line drawn through the tidal bulges does not point directly towards the Moon, but some distance ahead of it in the direction of the Earth's rotation. Instead of experiencing a high tide each time the Moon culminates in the sky, high tides at the Earth's coasts precede the Moon by several hours. Low tides take place during the intervals between the arrival of high tides, but local topography usually dictates that they do not happen precisely midway between high tides.

One of the most startling consequences of the offset tidal bulge is that it gravitationally tugs the Moon along in its orbit, increasing the Moon's orbital velocity and producing a slow recession from the Earth. Every century the Moon moves away from the Earth by 3.8 metres – an extremely slow recession, but one that has actually been measured by shooting laser beams at reflectors left on the Moon by Apollo astronauts.

The Moon illusion

When the full or near-full Moon is close to the horizon it often appears unexpectedly large – 'as big as a plate' is a simile that has been used to describe this stunning effect. In reality the Moon's apparent diameter becomes increasingly *smaller* as it approaches the horizon, because the observer, situated on a revolving planet, is moving away from the Moon. While the phenomenon of apparent shrinkage is far too subtle to be detected without optical aid, naked-eye observation with a simple cross-staff will prove to anyone's satisfaction that the Moon remains more or less the same size whether it is riding high in the sky or hovering over the horizon.

The Moon's apparent diameter is always around half a degree, regardless of its location in the heavens. Nevertheless, the Moon illusion is powerful. It arises mainly because we perceive the sky as being shaped like the interior of a flattened dome – an illusion reinforced by the perspective effect of clouds seen during the daytime, passing by underneath this dome. Celestial objects appear to be attached to the dome's interior. When the near-horizon Moon is viewed we imagine that it is far away and must therefore be a large object to appear the size it does. When the Moon is above us we subconsciously imagine it to be closer to us and smaller in size. Exactly the same illusion occurs in our perception of the size of constellations. Castor and Pollux, for example, seem more widely separated when they are near the horizon.

The plane of the Moon's orbit around the Earth is inclined to the ecliptic, the plane of the Earth's orbit around the Sun, by a little

◄ *The Earth and Moon are compared in this composite of photographs taken by the Galileo spaceprobe en route to Jupiter in 1992. The respective sizes of both objects are accurate, but the brightness of the Moon has been exaggerated.*

over 5°. The two points at which these planes intersect are called the ascending and descending nodes (respectively, where the Moon passes from south of the ecliptic to the north, and vice versa). Whether a solar or lunar eclipse occurs at new or full Moon depends on the position of the nodes in relation to the Sun and the Earth. The imaginary line connecting the ascending and descending nodes gradually moves from east to west, and makes a complete rotation every 18.61 years. So, every 18 years or so, from northern temperate regions such as the UK and much of the continental USA, on those occasions when the full Moon is at its lowest point in the zodiac in the constellation of Sagittarius and at its maximum apparent distance south of the ecliptic, it appears barely to clear the southern horizon at midnight at culmination. From New York (41°N) the Moon rises 20° above the horizon. From London and Calgary (52°N) the Moon rises 9° above the horizon. From Edinburgh (56°N) this is reduced to just 5°, while from Lerwick in the Shetland Islands (60°N) the full Moon appears to physically roll along the North Sea horizon for under two hours – a spectacular sight. This phenomenon is next due to occur around midnight on 21/22 June 2024.

From southern temperate latitudes, the full Moon appears very low above the northern horizon during December. From Alice Springs (24°S) the lowest midsummer full Moon rises as high as 38° above the northern horizon. From Melbourne (37.5°S) the full Moon culminates at a somewhat lower altitude, at 24.5°. However, there are no mainland locations anywhere near the 60°S line of latitude from where the Moon will appear to roll along the horizon, as it appears to do in northern midsummer.

Following the Moon in the sky

As it moves from west to east in the sky, the Moon travels an angular distance a little more than its own diameter every hour. In 29.5 days it goes through a complete set of phases, from new Moon, through full Moon and back to new Moon again; this period is called a synodic month, or lunation. Because the plane of the Moon's orbit lies close to the ecliptic, its monthly path among the constellations of the zodiac is similar to that followed by the Sun during a whole year.

The full Moon is, of course, always on the opposite side of the sky to the Sun. At the winter solstice, when the Sun is at its lowest point above the horizon, the midwinter full Moon rides high in the early morning skies. At the summer solstice the Sun reaches its highest point above the horizon, whereas a midsummer full Moon just manages to

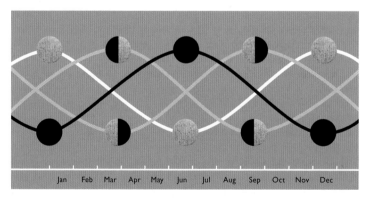

▲ Each month the Moon follows an orbit close to the plane of the ecliptic, and its height above the horizon varies, just as the Sun varies in altitude through the year. This graph shows the varying altitude of the Moon for its four main phases at southern culmination from northern temperate regions. For example, winter full Moons are high, while summer full Moons barely nudge above the southern horizon at midnight.

pull itself clear of the horizon for a few hours. Around the time of full Moon in the northern-hemisphere autumn, the ecliptic (and therefore the Moon's orbital plane) makes a very shallow angle with the eastern horizon, causing the Moon to rise only a few minutes later each evening. The Harvest Moon is the full Moon nearest the date of autumnal equinox in the northern hemisphere, on or around 23 September. In the northern-hemisphere spring, the ecliptic makes a steep angle with the eastern horizon which means that around the time of full Moon its consecutive evening rising times are widely separated – sometimes by as much as an hour and a half. Observers in the southern hemisphere experience a Harvest Moon of sorts during southern-hemisphere autumn in March, when consecutive rising times of the full Moon are closest together. However, the separation between the Australian Harvest Moon's rising times is longer (even from far southern Australia at around 40°S) than it is from the UK or North America (at latitudes greater than 50°N) because of the slightly steeper angle that the ecliptic makes with the horizon.

Lunar librations

As the Moon orbits the Earth it keeps the same face – the nearside – turned in our direction. Because of a phenomenon known as libration – an apparent back-and-forth rocking motion of the Moon's globe which is far too slow to be noticeable in real time during an observing session – means that from the Earth a total of 59% of the Moon's surface can theoretically be observed at one time or another. The remaining 41% constitutes the permanent lunar farside, forever hidden from the terrestrial observer. The regions bordering the 90°E and 90°W lines of longitude are known as the libration zones, and features occupying these zones are often librated past the Moon's limb, making them unobservable from the Earth.

There are two fundamentally different types of libration – optical and physical. Optical libration makes by far the greatest difference to what is observable, and results from the ever-changing presentation of the Moon's face to the terrestrial observer, combined with the position on the Earth from which the Moon is viewed. Physical libration is the wobble of the Moon about its own centre of gravity, and it is an exceedingly small oscillation. The most obvious effect of optical libration is on the apparent position of features near the limbs. There are three kinds: libration in longitude, libration in latitude and diurnal libration.

Libration in longitude is a consequence of the Moon's angular velocity varying as it follows its elliptical orbit around the Earth, but maintaining a constant rate of rotation upon its own axis. When the Moon's rate of axial rotation and its orbital progress about the Earth are 'out of phase', observers can see farther round the eastern or western limb. If the illumination is right, telescopic observers will take the opportunity to record the appearance of features thus brought into view. Every anomalistic month (the lunar month measured from perigee to perigee), libration in longitude favours the eastern limb, followed by a favourable libration of the western limb a fortnight later, displacing the centre of the lunar disk by +7° 54′ west and east, respectively.

The next type of libration arises because the Moon's axis of rotation is not perpendicular to its orbital plane around the Earth: the lunar poles are actually tilted slightly. This enables the observer, in each anomalistic month, to peer a little farther over the northern limb, and a fortnight later to see a little farther under the southern limb. Libration in latitude, as this phenomenon is called, displaces the mean centre of the Moon's disk to the north, and a fortnight later to the south, by +6° 51′.

In addition to libration in longitude and latitude, the observer's view of the Moon changes because the Earth itself is rotating, carrying the observer around with it in the course of a night. The extent of this so-called 'diurnal' libration depends on the observer's terrestrial location on the globe; the greatest amount of diurnal libration is observable from equatorial regions where more of the Moon's eastern or western limb can be seen at moonrise and moonset (about one degree extra on either side).

The three causes of optical libration combine to give rise to a libration zone, consisting of 18% of the Moon's surface area. The visible part of this zone moves clockwise around its limb once a month, though the exact areas which come into view vary from one lunation

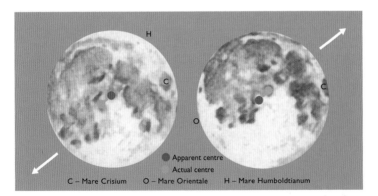

C – Mare Crisium   O – Mare Orientale   H – Mare Humboldtianum

▲ A comparison between two extremes of libration, showing the most favoured direction of libration and the offsetting of the centre of the Moon's disk. At left, libration favours the north-eastern limb, and Mare Humboldtianum is very favourably placed for observation. At right there is a favourable south-western libration, with Mare Orientale on display.

to another. Over time, all the features within the Moon's libration zone eventually make an appearance on the lunar limb. A narrow crescent-shaped segment of the libration zone is presented to the Earth at the vast majority of times, though this region is not always visible because it might be in shadow. It is strange to think that the 'mean' Moon – one which has the 90° longitude line positioned exactly around the visible limb, as depicted on most lunar charts and atlases – is rarely what we see.

Lunar observers take optical librations into account because they determine the practical visibility of features near the Moon's limb. Too much foreshortening of features near the limb prevents useful observation. For example, it is no good planning in advance to observe Mare Orientale, which lies largely past the 90°W line near the Moon's south-western limb, without knowing whether libration will have brought it into view. Similarly, it is best to plan to observe nearside limb features when they are presented favourably. There is little point, say, in attempting to observe the floor of the walled plain Gauss, which lies near the north-eastern nearside limb, when there is a strong libration favouring the south-western lunar limb.

Libration occasionally brings into view areas of the north and south lunar polar regions that are observationally challenging because of the great profusion of foreshortened craters at the Moon's edge. It has been speculated that some deep, permanently shadow-filled craters near the Moon's south pole harbour water-ice deposits mixed in with the lunar soil. Although it isn't possible to directly view the interiors of these craters because their floors are permanently veiled with shadow, the elongated rims of many of them can easily be seen during a good libration.

Features closer to the centre of the lunar disk are affected not so much by alteration in apparent shape but by their proximity to the Moon's terminator, hence their individual times of emergence from and immersion into the lunar night. Numerous astronomical almanacs contain tables for the calculation of libration. Many good astronomical programs available for personal computers are capable of displaying libration and the lunar phase graphically, taking out the time-consuming mathematics involved in calculating and plotting an 'individual' terminator.

The combined effects of libration in all its forms have profound implications for the telescopic lunar observer. When seen from the surface of the Moon, the Earth hangs forever in the same region of the sky, slowly migrating around this part of the sky in response to the Moon's librations. Picture the scene in the middle of the next century – hundreds of small dish antennae across the nearside lunar surface, all making their own little synchronized librations to maintain their alignment with the Earth.

Earthshine

Earthshine is the faint illumination of the Moon's dark side, caused by sunlight being reflected on to the Moon by the Earth. The phenomenon is most obvious when the Moon is a thin waxing crescent in a reasonably dark sky – a sight that was sometimes referred to as 'the old Moon in the young Moon's arms'. Viewed from the night side of the Moon's surface at this time, the Earth would appear as a bright waning gibbous sphere lighting the lunar landscape with up to 60 times as much brilliance as the full Moon seen from the Earth.

With the naked eye, earthshine is visible for a few days after new Moon, and picked up again at the end the lunar month. The brightness of earthshine actually varies according to Earth's phase, geography and global weather conditions – continents are more reflective than oceans, and a cloudy globe is highly reflective. Anyone who regularly observes the crescent Moon will soon come to notice that there are times when earthshine seems especially prominent.

Moonlight

While viewing a brilliant full Moon high in the sky, it is difficult to appreciate that our satellite is one of the least reflective worlds in the entire Solar System. It has an average albedo (reflectivity) of just 0.07 – only 7 out of every 100 photons that hit the lunar surface bounce back into space. The albedo range of the lunar maria varies between 0.05 and 0.08, while the brighter highlands have an albedo range from 0.09 to 0.15. The intensity of moonlight is only 0.25 lux – a quarter the intensity of a burning 'standard' candle placed a metre away. If the Moon were a smooth sphere, its average albedo would be a little higher. But its surface is rough and irregular, and largely because of the shadows thrown up by these features, the brightness of the half Moon at first or last quarter is not half of the full Moon's value as determined by its area, but just one-ninth. As bright as the full Moon appears, sunlight is half a million times brighter. A terrestrial landscape bathed in moonlight has a ghostly monochrome appearance because the scene is not bright enough to trigger all of the colour receptors in the human eye.

The next time you catch sight of the Moon, pause for a while to consider the Earth's only natural satellite – its size, distance and orbit – and reflect that human culture might have been incalculably poorer had the Earth not been accompanied by such a big, bright satellite. Indeed, life itself may not have got started at all without the Moon's gravity and the ocean tides it causes.

THE LUNAR OBSERVER'S EQUIPMENT

Your eyes are by far the most important optical equipment you will ever possess. Look after them properly, and you will be able to enjoy the full majesty of the lunar surface for as long as you wish to carry on observing. Despite its apparent brightness through the eyepiece, the concentrated light of the Moon cannot damage your eyes in any way. Moon filters that screw into the barrel of the eyepiece simply reduce glare and enhance contrast – they are not intended to prevent eye damage, in the way that solar filters are (see Chapter 6).

Many observers experience floaters – tiny specks in a variety of shapes and sizes that become visible in one's field of view when observing a bright object such as the Moon. Floaters are not necessarily a sign of unhealthy eyes. They are actually the shadows cast on to the retina by the remnants of dead cells floating in the vitreous humour, a gel found within the main compartment of the eye. Annoying though they are, one has to put up with them, and they tend to increase in number with age. In extreme cases eye surgery can remove floaters, but this is a last resort for people who are severely debilitated by the condition, and there are of course risks associated with any surgery. Annual eye check-ups are recommended, as some unsuspected but treatable medical conditions may come to light as a result.

Binoculars

Viewed through the most modest of optical equipment, the Moon's surface opens into a remarkable vista of seas, mountains and craters by the score. Even a pair of small and unsophisticated opera glasses (which have the same basic optical configuration as Galileo's trusty old telescope) will magnify the Moon sufficiently to show it as a rugged, crater-scarred globe.

In this world of high-tech computer-driven instrumentation, many observers remain quite happy to use nothing grander than a steadily held pair of binoculars in their pursuit of the Moon's wonders – and why not? On the whole, binoculars have the advantage of being less expensive, far easier to transport and to use, more able to withstand everyday accidental knocks and far more useful for non-astronomical purposes than telescopes. Provided they are of sound optical quality, any pair of binoculars will show enough detail for the observer to follow the changing appearance of the larger features as they emerge into the lunar sunshine along the terminator. Binoculars with a variable low to high power range allow the observer to zoom in for closer scrutiny.

The Moon passes by some lovely star groupings during the lunar month, and binoculars are the best way to see the Moon in a low-power wide field. And the use of both eyes enables the observer to gain the maximum benefit from viewing in (apparently) three dimensions. The use of a simple steadying rod or a lightweight photographic tripod is advised – it will increase the effectiveness of any binoculars and enhance anyone's enjoyment of Moonwatching. Some high-end binoculars have built-in image stabilizers which compensate for the effects of shake – they are amazing to use, but very pricey.

You can tell from the wide variety of binocular shapes and sizes on sale at your local store that a number of different optical configurations are available, which largely determine how much detail you can see and the overall quality of the view. As with anything, you get what you pay for (provided you shop with a reputable optical dealer), but it is possible to buy a good pair of binoculars for under £50 or US$80.

Galilean binoculars (also known as opera glasses) are the simplest and cheapest optical instruments on general sale today. They utilize biconvex objective lenses and biconcave eye lenses to form an upright image. With their short focal length they offer low magnification and a narrow field of view, and chromatic aberration (see page 50) is quite evident. Compact and lightweight, Galilean binoculars can be carried in your pocket or bag, and they can be used to see which features are visible along the Moon's terminator, enabling the observer to decide whether to set up the telescope to scrutinize these features in greater detail.

Serious binoculars use prisms to fold the light path in order to reduce the length of the binoculars and turn the image the right way up for terrestrial viewing. Prism binoculars are optically far superior to Galilean binoculars and offer a far wider field of view and higher magnification. Their power is identified by two figures denoting the magnification and the aperture of the objective lenses: for example, 10×50 binoculars give a magnification of $10\times$ and have 50 mm objective lenses. For general stargazing 7×50 or 10×50 binoculars are ideal, since they deliver a wide field of view, a bright image and a magnification low enough to be able to scan the skies for short periods without the need to support the binoculars. A pair of 7×50s gives a field of view of about $7°$ across – 14 times the apparent diameter of the Moon. Naturally, only so much can be viewed on the lunar surface at such a low magnification, but the surrounding starfields (including near passages of the brighter planets) can be delightful, and lunar occultations of the brighter stars can be observed (see Chapter 6).

To see the Moon in more detail you will need higher-power binoculars, but they have a narrower field of view, and shake is more of a problem as magnification is increased. It is virtually impossible to enjoy a view of the Moon with hand-held binoculars at magnifications greater than $10\times$ – you need support or a tripod to keep them steady. A magnification of $20\times$ or greater will enable a great amount of lunar

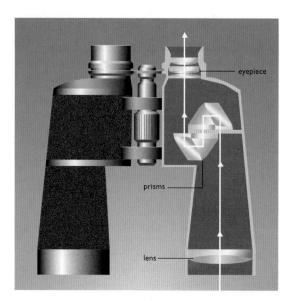

◄ A schematic cutaway illustrating the optical layout in Porro prism binoculars. The prisms allow portable, compact binocular designs.

eyepiece

prisms

lens

surface detail to be discerned – a pair of 20 × 80 binoculars will deliver truly stunning, almost three-dimensional lunar views. Zoom binoculars that can be adjusted to magnify between, say, 20× and 100×, are great for lunar sightseeing, but they are less effective for general stargazing than fixed-magnification binoculars as there is some light loss. Additionally, the mechanism required to move the lenses internally may affect image quality to a certain degree because the lenses require excellent alignment throughout their range of focus, and the quality of construction is sometimes wanting in cheaper zoom models.

Two types of prism binoculars are available: Porro prism binoculars and roof prism binoculars. Porro prism binoculars are the most widely used type today. Traditional designs have a 'W' shape where the prisms fold and offset the light path from the objectives to the observer's eyes. In recent years the styling of small Porro prism binoculars has changed, and many are now encased in a U-shaped shell with objective lenses that may actually be closer together than the eye lenses – the result of an inverted Porro prism design. Roof prism binoculars are becoming increasingly popular because of their compactness and light weight. 'Roof' is a reference to the shape of the prism, and they are sometimes called 'Dach' prism binoculars (*Dach* is German for 'roof'). Most roof prism binoculars have a 'straight through' shape that belies their power, and are very compact. Finally, half a binocular is better than none at all. Monoculars are inexpensive little roof prism telescopes that deliver lovely low-power views – ideal for casual Moon-spotting and viewing nearby bright stars and planets in the same field of view.

Telescopes of all shapes and sizes

There are few sights in astronomy as stimulating and as absorbing as a high-magnification view of the Moon through a telescope. With its surface smeared by dark lava flows, punctuated by mountain ranges, pockmarked with craters, and raked with rilles and ridges, the Moon is the most popular astronomical object for amateur observers. Its grandeur has lured every single amateur astronomer armed with a telescope into gazing at the lunar landscape, and for many the love affair with our satellite lasts a lifetime. Anyone can browse the Internet or flick through the pages of a book to view high-resolution images of the Moon secured from big telescopes on the Earth and from spacecraft in lunar orbit. But nothing compares with seeing the Moon for real through a telescope.

Small scopes The Moon is an impressive sight which, in the four centuries since Galileo first scanned the lunar landscape with his tiny 16 mm refractor, few amateur astronomers have ever tired of. It is not surprising that a great many star parties and public outreach events organized by astronomical societies are timed to coincide with an evening crescent Moon – if the skies are clear, our natural satellite invariably becomes the event's star attraction.

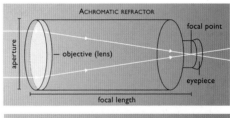

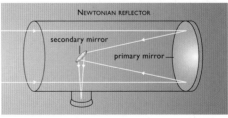

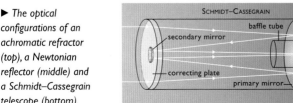

▶ *The optical configurations of an achromatic refractor (top), a Newtonian reflector (middle) and a Schmidt–Cassegrain telescope (bottom).*

However large or attractive a telescope may look, its real value lies in its optical quality. Avoid inexpensive telescopes advertised in newspapers with promises of 'See the stunning rings of Saturn! Marvel at a multitude of Moon craters!'. More often than not, such instruments are constructed entirely of plastic, including the lenses, which can perform very poorly indeed. They are a complete waste of money and moonlight, a lesson in how an astronomical telescope should not be, and are fit only for theatre props or children's toys – provided the children know that real telescopes are much better.

Small telescopes in the windows of department stores are of variable optical quality – usually somewhat less than mediocre – and over-priced. If you are intent on spending your money here, then you are entitled to take the instrument out of the box and examine the optics for flaws and scratches, and then to test the view through the eyepiece (preferably by focusing on a distant object outside the store). A store that refuses to allow you to make this basic inspection is clearly not interested in your custom. Make sure you buy your first telescope from a reputable company that specializes in optical or astronomical equipment. Such companies advertise extensively in astronomical magazines, and most publish a catalogue online and/or in print.

▼ *The author's computer-controlled 200 mm Schmidt-Cassegrain (left) and manually-controlled 300 mm Newtonian telescope (right).*

Some observers believe that any telescope with an aperture smaller than 75 mm is a mere toy, incapable of providing any meaningful views of the lunar surface. I could not disagree more. I can vouch that examples of just about every type of lunar feature can be seen through a telescope as small as a 40 mm refractor of good optical quality, including maria, mountains, hundreds of craters, some of the larger faults, valleys, wrinkle ridges and domes. It is true, however, that a small telescope will not reveal the finer lunar detail, and the effective maximum magnification of a 40 mm refractor is 80×.

Because they are lightweight and highly portable, small telescopes are perfect for casual lunar sightseeing when the effort to set up a larger telescope seems too daunting; they are perfect for use on nights of mediocre seeing, to catch views of the Moon in breaks between cloud. Paradoxically, during periods of poor seeing a small telescope may sometimes appear to give a sharper, more shimmer-free image than a larger instrument used at the same magnification. This is because the image formed by a smaller instrument is not as prone to degradation by atmospheric turbulence as is the image formed by a larger one.

Many amateur astronomers begin by buying modest equipment, upgrading through the years as budgets permit. It is a common aspiration to own a computer-guided instrument of 200 mm aperture or greater to satisfy more advanced requirements. But very small telescopes have a number of advantages. They are lightweight, highly portable, and to some extent 'expendable' – accidental damage to the optics or mechanics of a small budget scope can be written off as 'just one of those things'. Some of the cheaper small telescopes have a fixed magnification and the eyepiece is non-interchangeable. Others have a built-in variable-focus eyepiece; being able to switch between low, medium and high magnification with ease can be a great boon in lunar observing. The best small telescopes come with a selection of two or three eyepieces, giving sensible magnifications that do not overstretch the instrument's capabilities, and these are often of the 'Japanese inch' variety, with a 0.965-inch barrel diameter.

While some of the cheaper telescopes have acceptable achromatic (free of false colour) objective lenses, the eyepieces may be of the most rudimentary nature – simple Huygenian eyepieces that present claustrophobic tunnel views with an apparent field of view of 30° or even less. If you have a small scope with 0.965-inch (24.5 mm) eyepieces, it is worth investing in one or two good-quality 1.25-inch (32 mm) Plössl eyepieces that give a decent field of view of about 50°, and a 0.96-inch/1.25-inch eyepiece adapter (these adapters can be bought as star diagonals). By using good-quality eyepieces you can gauge the real optical performance of your small telescope, and if you later decide to upgrade to a larger instrument you will have two eyepieces that you can use with it. After upgrading you might also want to mount your small telescope on top of the new instrument in order to use it as a finder or guide scope – ideal for getting the Moon adequately centred in the field of view when imaging the Moon using a compact camera afocally or with a webcam. (For more on eyepieces, see pages 56–60.)

'Serious' telescopes

An increasingly wide range of high-quality telescopes is available to the amateur astronomer, advertised widely in the major astronomy magazines. To the beginner, the variety of instruments on offer can be overwhelming. How do complete novices go about choosing the telescope that is right for their needs? Is there a 'right' telescope for lunar observing?

Refractors Refracting telescopes consist of a tube with an objective lens at one end and an eyepiece at the other. The objective lens collects and focuses the light, and the eyepiece magnifies the focused image. Telescopes are often characterized by their f number, or focal ratio. A refractor described as being f/10 may have a 100 mm diameter objective lens with a focal length of 1000 mm. Refractors are most people's idea of what a 'proper' telescope ought to look like.

The English amateur astronomer Thomas Harriot used a small, very basic refractor to make the first-ever telescopic drawing of the Moon, on 5 August 1609, beating Galileo by several months. The earliest refractors consisted of a single objective lens and a single eyepiece lens, and they usually had very long focal lengths to combat the effects of false colour, technically known as chromatic aberration. Red light is brought to a focus farther from the objective lens than blue light, and no matter how much the observer twiddles the focuser, there will always be false colour in the image. The fatter the lens, the shorter its focal length and the greater the difference in focus between red and blue light will be.

In today's decent-quality astronomical refractors the objective lens comprises two specially shaped elements positioned in very close prox-imity, the combination being chosen to focus all the different wave-lengths of light as near to a single point as possible. Such achromatic objectives reduce (though they do not absolutely eliminate) chromatic aberration, and it is noticeable in, for example, budget short-focus refractors produced in China. According to some optical purists, such short-focus (f/8 and below) achromatic refractors are not capable of delivering an image that the amateur astronomer would consider satis-factorily free of false colour. But they do perform surprisingly well, and though they display false colour around the edges of bright objects such as the Moon and planets, this is restricted to a violet tinge that is aesthetically unobtrusive. An inexpensive way to eliminate most of the violet tinge is to use a minus-violet contrast-boosting filter. Alternatively, a special lens called a Chromacorr can be fitted into the focuser to correct for the colour aberrations optically, rather than by filtering, and although it is very expensive it effectively transforms the instrument into a true achromat. In all other respects the optical quality of the majority of these Chinese refractors is excellent, with good contrast and image definition across the field.

In recent decades a number of quality telescope manufacturers have introduced their own so-called apochromatic refractors – instruments capable of delivering images that are almost free of chromatic aberration. Apochromats use special glass or crystal in an objective lens consisting of two or three elements. Because of their near-perfect, high-contrast images, apochromats are one of the best kinds of telescope for lunar and planetary studies. Such performance does not come cheap – on average, apochromats are ten times more expensive than similar-sized achromats.

◀▲ The author's computer-controlled short-focal-length 100 mm achromatic refractor (left). This lightweight and portable instrument delivers lovely crisp views of the Moon and is an ideal 'grab and go' telescope. It is shown (above) set up for prime focus astrophotography with a DSLR.

The objective lens of a refractor will have been collimated (aligned) in the factory and secured in the tube, ready for immediate use once the telescope is taken out of its packaging. Refractors require occasional external cleaning, and dust particles should be carefully removed from the outsides of the lenses with a soft optical brush. Sometimes, for no reason other than curiosity, the telescope owner feels compelled to dismantle objective lenses or eyepieces. The altered configuration of the reassembled lenses may then give a disappointing image quality. Resist the urge!

Reflectors The optical solution to the chromatic aberration inherent in refractors came in the late 17th century with the invention of a telescope that had a mirror rather than a lens to collect and focus light. Because light is reflected by a mirror rather than refracted, it is not separated into its constituent colours, and the focused image is therefore free of chromatic aberration. In 1672 Isaac Newton hit upon the idea of using a concave primary mirror and a flat secondary mirror that reflected the light sideways through the side of the tube into the eyepiece. This classic Newtonian design remains highly popular today.

For many years the 150 mm $f/8$ Newtonian reflector was the commonest amateur telescope, capable of giving pleasing all-round celestial views, with a low-power eyepiece for deep-sky observation and higher powers for lunar and planetary studies. But for detailed high-magnification lunar observation, longer focal lengths are preferable. A well-collimated, long-focal-length Newtonian reflector with high-grade optics can deliver views that rival those through an apochromatic refractor.

Invented by the French priest Laurent Cassegrain in the same year as Newton invented his reflector, the Cassegrain reflector has a primary mirror with a hole at its centre, through which the light is reflected back by a convex secondary mirror higher up the tube. Cassegrains achieved popularity in the mid-19th century, but even at the height of their appeal they played third fiddle to Newtonians and achromatic refractors since they were prone to a number of troublesome optical aberrations. Typically made with focal lengths from $f/15$ to $f/25$, and good for high-power lunar studies, Cassegrains are more difficult to adjust and collimate than Newtonians.

To keep them in good working condition, reflectors require a good deal more care and attention than refractors. Because they are exposed to the open air, the thin reflective aluminium coating of the primary and secondary mirrors will gather dust and other specks of anonymous debris, spider's webs, and the like, and the coating will deteriorate rapidly if carelessly exposed to the elements. Shine a flashlight down the telescope tube on to the primary mirror and a coating of dust and debris will invariably be seen. The grubbier a mirror gets the more light

► *The author's home-made 300 mm f/4.5 Newtonian reflector, mounted in a Dobsonian (altazimuth) fashion. Used in conjunction with a long-focal-length wide-field eyepiece, the entire Moon can easily be accommodated in the field of view of this instrument.*

it will scatter, reducing contrast. Mirrors must be cleaned with great care so as not to drag debris across the surface. The looser debris can simply be blown away with a puffer or compressed air, and if necessary the surface can be physically cleaned using cotton wool and lens cleaning fluid or lens wipes – but very gently, in single strokes, and never reusing the same piece of cotton wool or lens wipe.

A Newtonian can be protected by covering the aperture with high-quality optically transparent film (of the same feel as thin cellophane) to seal the tube. I use a disk of this material on my 250 mm Newtonian whenever I think that it might spot with rain at some point during the observing session (the film comes in lengths for you to cut to your own specifications, so if the cover gets dirty you can simply cut out another disk). If mirrors are well protected when not in use they can be expected to last the best part of a decade before they need to be sent away for realuminizing.

Catadioptric telescopes Instruments that employ both mirrors and lenses to collect and focus light are known as catadioptric telescopes. The first proper catadioptric was invented by Bernhard Schmidt in the 1930s and built as an astronomical camera for surveying large areas of the sky. The Schmidt camera used a large spherical mirror of low focal ratio whose inherent spherical aberration (its inability to focus all the light it collects into one point) was corrected by a very thin, specially figured lens in front of the mirror, known as a corrector plate. The innovative design of the Schmidt camera was the ancestor of today's highly popular Schmidt–Cassegrain telescopes (SCTs), which as their name suggests are a hybrid of the Schmidt and Cassegrain configurations. SCTs were first produced commercially in the 1960s by Celestron in the USA. My own SCT was made by Criterion (a US company no longer in existence) and was previously owned and used by the veteran lunar observer Harold Hill.

SCTs have a relatively large central obstruction to accommodate the secondary mirror. The presence of this obstruction in the incoming light path produces a degree of diffraction which very slightly compromises the quality of the image when viewing detail on the Moon and planets, compared with a refractor of similar aperture. Nevertheless, a well-collimated SCT with good optics can produce excellent lunar images. Modern SCTs are designed to accommodate a myriad of accessories useful to the lunar observer – filter wheels, cameras, webcams, digicams, camcorders and CCD cameras. Many SCT users choose to employ a star diagonal for viewing comfort. This flips the Moon's image to produce a left–right mirror image, making it difficult to use a conventional Moon map to identify lunar features. However,

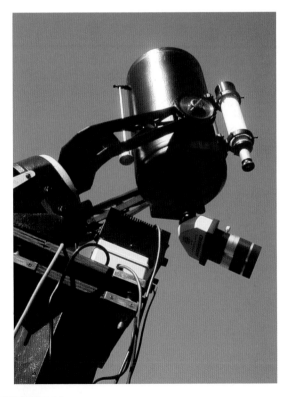

◀ The author's 200 mm Dynamax Schmidt–Cassegrain telescope. The instrument has a proud lunar observational history, having once been owned by the veteran lunar observer Harold Hill. It is shown here with its motor-drive unit and hand-controller, armed with a binocular viewer and variable-focus zoom eyepieces for stereo lunar observing.

▶ A view down the tube of the author's 300 mm Newtonian. Note the helical focuser, the primary mirror and the secondary mirror on its 'spider'. The entire tube can be rotated along its axis for ease of use.

reversed-image Moon maps are now available, aimed principally at SCT users. Schmidt–Newtonian telescopes have a glass corrector plate at the front end of the telescope and a parabolic primary mirror, but the flat secondary mirror reflects the light out of the side of the tube instead of channelling it through a hole in the main mirror.

Based on a design originated by Dmitri Maksutov in the 1940s, the Maksutov–Cassegrain telescope (MCT) has a spherical primary mirror and a deeply curved spherical meniscus lens at the front of the tube to correct for spherical aberration. The secondary mirror in the MCT is a small spot aluminized directly on to the interior of the meniscus. MCTs may resemble SCTs in many ways, but their performance on the Moon and planets far exceeds them. MCTs have longer focal lengths, and can give high-resolution, high-contrast views of the lunar surface with hardly any chromatic aberration. MCTs used to be specialist instruments, but they are now being mass produced and their price is very reasonable. My own 127 mm MCT is a budget Chinese instrument, but its performance is superior to my 150 mm achromat in many ways, especially in the lack of false colour. It is also very lightweight and portable, making it ideal for taking on observing trips to the country and on holidays.

The power to resolve

There is no shortage of fine lunar detail, and the ability to resolve it (assuming good seeing conditions) depends on the diameter of the telescope's objective lens or mirror. If the angle of illumination is right, a crater as small as 6 km can be recognized for what it is in a 60 mm refractor, while a 150 mm telescope will show a narrow rille just 150 metres across. The table should serve as a rough guide, though the constantly changing angle of illumination engages the Moon's relief features in a perpetual game of hide-and-seek.

THE SMALLEST LUNAR FEATURES VISIBLE IN TELESCOPES OF VARIOUS APERTURES			
Aperture (mm)	Max. magnification	Smallest crater (km)	Narrowest rille (metres)
Naked eye	(1)	200	—
40	80	12	500
60	120	6	350
100	200	3.5	250
150	300	2.5	150
200	400	1.8	110
250	500	1.4	90
300	600	1.2	70

Eyepieces

It goes without saying that an optically good eyepiece is as important to the performance of an instrument as its objective lens or mirror. The magnification provided by a given eyepiece depends on the telescope's focal length. To calculate this, simply divide the focal length of the telescope by the focal length of the eyepiece. For example, a 9 mm eyepiece will give a magnification of 133× in a telescope with a focal length of 1200 mm (1200/9 = 133).

Three good-quality eyepieces giving different magnifications should be enough to satisfy the basic needs of the lunar observer:

1 A low-power eyepiece with a wide field of view that can accommodate the whole lunar disk (0.5°) or a substantial portion of it. This eyepiece can be used for general celestial sightseeing, introducing friends to the Moon's surface, and for lunar eclipse observation.

2 An eyepiece of medium power (up to 100×) for more detailed lunar observation.

3 A high-power eyepiece. As a guide, the maximum magnification your telescope can bear is twice its aperture in millimetres, for example 120× for a 60 mm refractor. This eyepiece can be used if seeing conditions permit – when the Earth's atmosphere is not making the Moon's image wobble and shimmer – and fine lunar detail needs to be resolved.

Three eyepiece barrel diameters are on the market today – 0.965-inch, 1.25-inch and 2-inch. Plastic barrel 0.965-inch eyepieces are supplied with many of the cheaper smaller telescopes and are generally very basic two-element designs such as the Huygenian and Ramsden types, but sadly they can be of poor optical quality. Good 0.965-inch eyepieces used to be produced several decades ago by major companies like Zeiss and Vixen – for example, my own Vixen 60 mm refractor, purchased in the early 1980s, came with three very good 0.965-inch eyepieces. Good though a few of these 0.965-inch eyepieces may be, they look and feel insubstantial in comparison with larger designs. Most amateur astronomers have telescopes with focusers that accept 1.25-inch eyepieces exclusively, but an increasing number of focusers are fashioned with an adapter to accept both 1.25-inch and 2-inch diameter eyepieces.

Basic eyepieces Huygenian, Ramsden and Kellner eyepieces have very small apparent fields of view. They are not restricted exclusively to 0.965-inch barrels. All three sometimes masquerade under other names – such as 'modified achromat', 'super achromat', 'super modified achromatic' and 'super Ramsden' – to disguise their rudimentary nature to the prospective purchaser, so let the buyer beware!

The Huygenian design is one of the oldest types of eyepiece and consists of two plano-convex elements (the convex sides both facing the incoming light), and the focal plane lies between them. Huygenian eyepieces are suitable for use only with instruments of long focal length, $f/15$ or more, and give small apparent fields of 30° or less. Viewing the Moon through a short-focal-length Huygenian eyepiece can be like looking down a long tunnel – hardly a 'spacewalk' experience, and barely a spacecraft 'porthole' view! The small apparent field means that an undriven telescope requires constant adjustment to keep a particular feature near the centre of the field. Satisfactory lunar observing is a near-impossible task at magnifications over 100× with such an eyepiece.

The Ramsden eyepiece originates from the late 18th century and consists of two plano-convex lenses, both convex sides facing each other (the elements may in fact be cemented together to provide better correction), and the focal plane lies in front of the field lens. Ramsdens give flatter fields than Huygenians, but they produce more chromatic aberration. Their apparent field of view is restricted to around 30°, so they are best used for low-power views of the Moon.

The Kellner eyepiece was invented in the mid-19th century. It resembles the Ramsden, but the eye lens consists of an achromatic doublet. Kellners give better contrast but tend to produce annoying internal ghosting when bright objects like the Moon are viewed. Like the Ramsden, the Kellner's focal plane is in front of the field

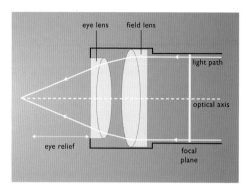

◄ A simplified cross-section through a Plössl eyepiece.

lens, so any tiny specks of dust that happen to land on the field lens are brought into view. Kellner eyepieces perform adequately, with good eye relief, if their focal length is more than 15 mm. Versions with shorter focal length produce blurriness around the edge of the field of view, and introduce noticeable chromatic effects – most unsatisfactory for observing the Moon at high power. Kellners have an apparent field of view of up to 50°, and are best used for low-to-medium-power lunar observing.

Better eyepieces Invented in the mid-19th century, orthoscopic eyepieces consist of four elements. Their name comes from their ability to produce a beautifully flat, aberration-free field, and they deliver very good, high-contrast views of the Moon. Their apparent field of view is around 50° and they have a reasonable degree of eye relief, giving a good degree of observing comfort. Many lunar and planetary observers swear by orthoscopic eyepieces, and I would concur that they are a useful accessory to have.

The Erfle eyepiece, invented in the early 20th century, is the first modern eyepiece design. With its arrangement of five lenses giving a low-power, 70° wide field of view with good colour correction, the Erfle was a real eye-opener to astronomers. Erfles perform at their best when used with long-focal-length telescopes, and the most effective versions are of 25 mm focal length and greater. Their edge-of-field definition, however, leaves something to be desired, and there is also the problem of internal reflections causing ghost images, which can be highly distracting when viewing the Moon at low power.

By far the most popular of today's eyepieces is the Plössl, though the design actually dates back to the mid-19th century. With their four-element design, modern Plössls have minimal internal reflections, good colour correction and an apparent field of view of around 50° which is flat and sharp right up to the edges. The Plössl's main disadvantage is its poor eye relief at lower focal lengths. For example,

I bought a 6.3 mm Plössl specifically for high-power lunar work, but it is virtually impossible to use effectively because you have to have your eyeball almost in contact with the eye lens to take in the whole field of view. Longer-focal-length Plössls give acceptable eye relief, and there are special long eye-relief versions with large eye lenses that allow spectacle wearers to observe in comfort.

Introduced in recent decades, modern wide-field and ultra-wide-field eyepieces such as the Meade Ultra Wide Angle (UWA) series, Celestron Axioms, Vixen Lanthanum Superwides, Tele Vue Radians, Panoptics and Naglers all deliver excellently corrected images with very large apparent fields of view. Larger fields of view are easier to use on an undriven telescope as the subject stays in the field for longer and you do not have to move the telescope as often. Naglers, with their 80° plus apparent fields, are splendid eyepieces that many amateur astronomers aspire to own. The longer-focal-length Naglers are big and heavy – the 13 mm Nagler is 135 mm long and weighs 0.7 kg! Changing between eyepieces of this sort and lighter regular eyepieces requires the observer to rebalance their telescope. Aside from this, the main downside of ultra-wide-field eyepieces is their price. A set of three or four Naglers can cost more than a new 200 mm SCT. Are they worth it? Of course, only the user can answer that, but suffice it to say that a view of the Moon through a Nagler is almost like viewing it from the porthole of a spacecraft in lunar orbit.

If you become tired of changing between eyepieces of different focal lengths, a zoom eyepiece may be just what you are looking for. Zoom eyepieces are by no means mere novelty items – several reputable companies sell a premium 8–24 mm zoom eyepiece. Their apparent field of view narrows from a generous 60° at 8 mm to a somewhat miserly 40° at 24 mm, but if you can accept this restriction, then a single zoom eyepiece can replace a drawer full of regular Plössls. They are great fun to use on the Moon.

Another fun item is the binocular viewer, a device that splits the single beam of light from the objective and channels it into two identical eyepieces. Because of their design, most will work only on a refractor or SCT – you may not be able to focus a binocular viewer at all using a Newtonian because they require a long light path. Binocular viewers can only be used with two identical eyepieces, and these ought to be of a focal length shorter than about 25 mm in order to prevent vignetting, an apparent fading effect that appears in the periphery of the field of view. I use two premium zoom eyepieces on my own binoviewer – it saves having to swap eyepieces to change magnification. Used on the Moon, a binocular viewer adds a sense of perspective to the lunar landscape. Even though you are viewing a two-dimensional image, the fact that both eyes are engaged gives it a near-3D quality, and viewing with both eyes enables finer detail to be discerned. The difference between observing with one eye and two can almost (but not quite) be compared with the difference between mono and stereo audio.

Telescope mounts

Any telescope needs to be mounted as rigidly as possible, for there is nothing more annoying than having the image bounce around the field of view at the slightest breeze. Undriven altazimuth mounts are the simplest type, and are often supplied in table-top form with small refractors. An altazimuth mount allows the telescope to be moved up and down and swung from side to side. The mechanical quality of many table-top altazimuth mounts leaves a lot to be desired. The bearings may be plastic, very small and adjustable by means of a simple friction knob. If the bearings are excessively slack then small pieces of card may be inserted between them to increase friction. If the mount is hopelessly shaky then it may be better to attach the telescope to a quality altazimuth camera tripod.

One type of altazimuth mount is the Dobsonian mount, popular because of its simplicity and ease of use. Dobsonians are used to mount reflectors of quite short focal ratio; they consist of a ground box that moves from side to side and another box holding the telescope tube that moves up and down. The use of low-friction materials such as polythene and Teflon for the load-bearing surfaces allows easy fingertip control, and lightweight structural materials such as plywood make Dobsonian-mounted instruments eminently portable.

A small telescope on a good undriven altazimuth or Dobsonian mount can be used quite easily with powers up to $100\times$. The higher the power, the faster the Moon's image will be carried across the field of view by the Earth's rotation, and the more frequently will small adjustments be needed to keep the Moon centred in the eyepiece. If the observer wishes to make an observational drawing, then there is little point in using magnifications above $100\times$, for during the briefest of glances at the sketchpad the subject of the drawing will be on its way out of the field of view. All but high-end altazimuth and Dobsonian mounts are rather uncooperative when it comes to making the very light adjustments needed to keep an object centred at high magnification. They often require quite a forceful nudge to overcome the friction on the bearings – completely removing the target from the field of view.

Observing the Moon at higher power demands a more substantial mount capable of following objects in the sky as the Earth rotates. When set up properly, an equatorial mount allows the observer to track celestial objects without the often frustrating pushing and shoving demanded by an altazimuth. Equatorial mounts have one axis that points towards the celestial pole, and another at right angles to it.

All that is needed to keep a celestial object centred in the eyepiece at high powers with a well-balanced undriven equatorial mount is either an occasional light push on the tube, a small twist to the slow motion knob or a touch on the electronic hand-controller.

There are two major types of equatorial mount in widespread use today. The German equatorial is a tried and trusted design. Highly adaptable, it can be used to mount refractors and reflectors. The entire sky, including the celestial pole, can be viewed with a German equatorially mounted telescope. Schmidt–Cassegrain telescopes are usually mounted on fork-type equatorial mounts, the telescope slung between the arms of the fork and the whole mount tilted to point towards the pole. A small region around the celestial pole is often out of reach of these instruments, since any large eyepieces or accessories such as focal reducers will not allow the telescope to swing fully between the fork and the base of the mount. This is no problem for the lunar observer.

Personal computers and lunar programs

Laptop computers are becoming a common sight alongside telescopes. Computers can guide telescopes and enable them to slew to a vast range of celestial objects. They can provide instant processing facilities for astronomical CCD imagery. The popularity of planetarium-type computer programs is steadily increasing. Most of them give basic lunar data such as the Moon's rising and setting times and its location in the sky. The Moon's phase is often expressed as a simple 'percentage illuminated' figure rather than days into the lunation, and while the phase depicted on screen may be spot-on, the graphic quality of the representation of lunar features can be variable and sometimes inaccurate. For casual observers, the basic lunar information contained in a planetarium program will more than suffice. Programs that enable the user to travel around the Solar System and view objects from any location are excellent educational tools. You can even stand on the virtual plains of Mare Tranquillitatis and see the sky as Neil Armstrong saw it back in July 1969, with a big waning Earth showing its Pacific hemisphere, almost directly overhead. Serious lunar observers are advised to shop around for a program that gives some or all of the following capabilities:

1 An accurate ephemeris of lunar data, including libration data and the Sun's selenographic colongitude.
2 A realistic zoomable picture of the Moon, showing the major features and taking into account the effects of libration.
3 The capability of predicting solar and lunar eclipses, with an animation of their progress.
4 The ability to show lunar occultations of stars and planets.

MOONWATCHING

As the Moon waxes, broadening from a narrow crescent through first quarter and gibbous phase to full, the morning terminator uncovers some of the Solar System's grandest topography. The Moon is one of the most geologically diverse bodies known, and boasts many easily observable examples of impact and volcanic cratering, multi-ringed asteroidal impact basins, ray systems, lava-flooded plains, lava flows, chains of secondary craters, faults, landslides, graben rilles, sinuous rilles, volcanic domes and wrinkle ridges. If you own or have access to a telescope, do spend some time each month lunar sightseeing – the Moon offers so much visual spectacle, and the show is absolutely free.

Although the Moon's ancient features are set in solid rock, its surface virtually indistinguishable from that perused by the earliest telescopic observers, the Moon is by no means a monotonous world. The continually changing play of light and shadow across its surface greatly alters the appearance of features from one day to the next. Incredible detail is brought into view along the morning terminator. Features that appeared mightily impressive on the terminator gradually become less grand as the lunation progresses. As the Sun climbs higher in the lunar sky, shadows shorten and sunlight washes features out, often fading them, sometimes banishing them from sight altogether. After full Moon, as the evening terminator crawls across the lunar disk, features fill with shadow once again, yet the same feature at sunrise and sunset can appear more different than might be expected. In addition, the Moon's continually shifting libration alters the appear-

LUNAR NOMENCLATURE		
With the exception of craters, lunar features have two-part names, the first of which is a 'descriptor' derived from Latin which identifies its type.		
Descriptor	**Meaning in Latin**	**Feature**
Catena	Chain	Chain of craters
Dorsum (pl. dorsa)	Back	Wrinkle ridge
Lacus	Lake	Small plain
Mare (pl. maria)	Sea	Large plain
Mons (pl. montes)	Mount	Mountain
Oceanus	Ocean	Very large plain
Palus	Swamp	Small plain
Planitia	Plain	Low plain
Promontorium	Promontory	Headland jutting into a mare
Rima (pl. rimae)	Crack	Rille (narrow valley)
Rupes	Cliff	Scarp
Sinus	Bay	Indentation at edge of mare
Vallis (pl. valles)	Valley	Large valley

ance of features near the edge of the lunar disk, sometimes allowing a glimpse of normally invisible terrain. The combination of changing illumination and libration means that the Moon never presents precisely the same appearance to an observer twice, even if that observer spends an entire lifetime viewing the Moon.

This chapter provides a detailed day-by-day guide to the Moon's features that are brought into view along the morning terminator during the first half of the lunar month, up to full Moon. Features are described from north to south. Mention is made of what can be seen with the unaided eye and binoculars, but most of the descriptions are of features visible through small telescopes (upwards of 60 mm). Where appropriate, the smallest aperture required to resolve a feature is given. Most of the features named in this tour of the Moon's surface are indicated on the accompanying maps. After full Moon the same features are uncovered by the evening terminator, in reverse order, and a more general description is provided for the second half of the lunation. The table opposite lists the so-called descriptors used in the official 'binomial' names of lunar surface features.

Observing the sub-24-hour lunar crescent

Naked-eye sightings of the very young crescent Moon are exceedingly rare. A one-day-old Moon is only around 13° from the Sun, and the glare of the dusk skies makes it almost impossible to catch sight of without optical aid. Binoculars are ideal for sweeping the western horizon after sunset in search of the very young lunar crescent. Because the Moon is not a smooth sphere, the young crescent never extends to a full semicircle around the limb – the shaded areas of high relief on the Moon's terminator are turned towards the observer and block out some of the sunlit portions behind them, and this effect is more pronounced towards the narrow horns of the crescent. Theoretically, however, a Moon just 14.5 hours old at an angular distance of 7.6° from the Sun can be observed as a crescent through binoculars.

During northern-hemisphere spring (March and April) the ecliptic makes its steepest angle with the western horizon, and this is the best time of the year to hunt for a very young crescent, since the Moon (always close to the ecliptic) will be at its highest above the horizon after sunset. In the far southern hemisphere, observers have a better view of the young crescent Moon during September and October, when the ecliptic makes a steeper angle to the western horizon after sunset. The higher the Moon is in the sky, the less atmospheric murk it has to shine through, increasing the observer's chances of seeing it. Viewed through a telescope, the day-old Moon presents a dim, irregular, wire-thin crescent shimmering with atmospheric turbulence – even the most experienced lunar observer would have difficulty in positively identifying any lunar features under these circumstances.

Day two

DAY 2

The 48-hour-old Moon is far easier than the day-old crescent to locate in darkening sunset skies. Against a fairly dark sky background, the dark side of the Moon can be seen, faintly illuminated by sunlight reflected from the Earth. Through binoculars the appearance of the blue-tinted earthshine-lit hemisphere being cradled by the bright directly illuminated crescent is quite beautiful, and the observer may be struck by a distinct three-dimensional impression of the lunar globe floating in space.

The most obvious feature of this young crescent Moon is Mare Crisium (Sea of Crises – 620 × 570 km) which is bisected by the morning terminator. Through binoculars or a 40 mm telescope at low power, Mare Crisium looks like a giant 'regular' crater – one so huge that it utterly dominates the scene. At low

magnification, when the eastern mountain border of Mare Crisium has just emerged into the morning Sun, broad black shadows are cast on to the dark, barely illuminated plain of the mare itself. Keen-sighted observers can see the apparent 'dent' in the terminator caused by Mare Crisium without any optical aid.

Through a telescope, a number of prominent wrinkle ridges can be traced across the surface of Mare Crisium, notably Dorsa Tetyaev (150 km long) and Dorsa Harker (200 km long) in the eastern part of the mare, which sweep in arcs north–south across the plain. Mare Crisium's north-eastern mountain ramparts give way to the sinuous lava-filled valley of tiny Mare Anguis (130 × 30 km) – the 'Serpent Sea', the smallest named mare on the Moon. Promontorium Agarum, a broad, square-cut mountainous cape, is a prominent feature on the eastern border of Mare Crisium, and casts a dark shadow eastwards on to the mare.

MARE FECUNDITATIS

MARE SMYTHII

Langrenus

Lohse · Kapteyn

Lamé

Vendelinus

— Holden

—Wrottesley

Petavius

Palitzsch

Humboldt

Snellius

Furnerius

Fraunhofer

MARE AUSTRALE

Pontécoulant

▲ The two-day-old Moon (58 hours) in a photograph taken on 25 March 1993 by Paul Stephens, using a 300 mm Newtonian.

About 50 km to its south lies Mons Usov, 15 km long and 1000 metres high, though it is rather an insignificant mountain in lunar terms. A hundred kilometres south-east of Mare Crisium, the irregular dark lava plains of Mare Undarum (Sea of Waves – 243 km) and Mare Spumans (Foaming Sea – 139 km) mingle with numerous other dark, flat-floored craters such as Firmicus (56 km), Apollonius (52 km) and Dubiago (46 km).

Most of the craters visible on the two-day-old crescent are at least partially filled with shadow, depending on the depth of their floor beneath their rim and their distance from the morning terminator. Because the observer is viewing directly into the shadowed parts of these craters, many of the subtle tonal variations of the easternmost maria (described below) and the craters in the east are not yet discernible because their eastern walls are casting internal shadows. However, as the lunation progresses and the Sun climbs higher, the patchwork of albedo features (albedo is the true reflectivity of the surface) becomes more evident, enabling the observer to differentiate between the dark lava-filled lowlands and the brighter highlands.

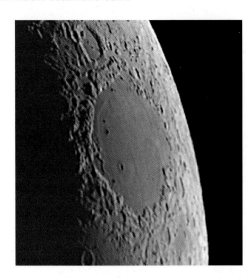

◄ *Mare Crisium in the early morning. This CCD image was obtained by Cliff Meredith, using a 200 mm SCT and a MX5 CCD camera.*

A couple of hundred kilometres south of Mare Crisium, another large lunar sea, Mare Fecunditatis (Sea of Fertility), is beginning to emerge from the morning terminator, along with its little positive-thinking bay, the Sinus Successus (Bay of Success) in the north-east. Nearby, the inconspicuous flat-floored crater Webb (22 km) appears slightly polygonal in outline. Just off the eastern border of Mare Fecunditatis lies Langrenus (132 km) with its impressive central mountains, broad terraced ramparts, and extensive system of external ridges and furrows. Roaming among the craters along the terminator, the eye is drawn to Langrenus with its bright inner walls. Farther south along the terminator, two more large craters are coming into view. Vendelinus (147 km), 100 km south of Langrenus, has considerably eroded walls that have been dented by subsequent impacts. Its flattish floor contains several small craters, and larger craters adjoin its eastern wall, including Holden (47 km) and Lamé (84 km). Lamé's formation took place long after Vendelinus was formed, and the impact that created it produced a series of high ridges on Vendelinus' floor.

About 100 km south of Vendelinus lies the magnificent crater Petavius (177 km) – larger than Langrenus and every bit as impressive when viewed at high magnifications. Petavius has an extremely compli-cated terraced wall, and like Langrenus its outer flanks are striated with radial ridges and furrows, traceable to distances of 150 km or more. As the morning shadows shorten, Petavius' incredible floor is uncovered. A sprawling mountain massif dominates the centre of the floor, with peaks that rise to more than 1500 metres. Several rilles extend across the floor. The largest of them, Rimae Petavius – visible in a 60 mm refrac-tor, runs fairly straight south-west from the base of the central mountains for a distance of 60 km to the base of Petavius' south-western inner wall, where it then veers to the south at a near right angle. This particular rille is a fault feature – a graben caused by crustal tension. The other rilles on Petavius' floor are sinuous and appear to have been caused by the outflowing of very runny molten lava, carving its way into the lunar surface by erosion, like a river valley. One of these sinuous rilles winds its way north from the central mountain (where the lava flow had its origin) and gradually diminishes in width and depth as it approaches the inner northern wall. There is another, narrower sinuous rille on the eastern part of the floor, but this requires a 150 mm aperture to resolve. Low ridges and lava stains occupy the rest of Petavius' floor. Adjoining Petavius' western wall is the crater Wrottesley (57 km), appearing somewhat deeper than Petavius, complete with its own central mountains. On the opposite wall, the irregular crater Palitzsch (41 km) elongates northwards, extending into a wide valley parallel to Petavius' rim for a further 110 km, where it is known as the Vallis Palitzsch. All these features, illuminated by a morning Sun, make an impressive sight through the eyepiece of a small or large telescope.

Several other large craters are visible on the post-two-day-old lunar crescent. Gauss (177 km), near the north-eastern limb, and Humboldt (207 km), towards the south, are large, very ancient walled plains with relatively low, uncomplicated rounded walls, and both features have broad floors which are crossed by systems of narrow rilles, mountain chains and craters, though their proximity to the limb makes them extremely foreshortened. Like the other features near the eastern lunar limb, they are best observed shortly after full Moon, illuminated by an evening Sun, when libration is favourable.

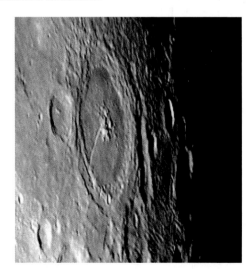

▶ Petavius, with its large internal rille. The image was obtained on 1 December 2001 by Mike Brown, using a 370 mm reflector and a HX516 CCD camera.

Day three

The whole of Mare Crisium is now visible, and the earthshine remains visible with the naked eye, faintly illuminating the dark side of the Moon. Through binoculars, all the major maria can be made out in the earthshine, plus several of the brighter regions. As features near the eastern limb begin to lose their internal shadow, binoculars start to reveal a number of near-limb marial areas clearly, depending on libration. The dark patch of Mare Humboldtianum (160 km) can be made out near the north-eastern limb, and at a favourable libration the irregular grey patches of Mare Marginis (Border Sea – 150 × 580 km) and Mare Smythii (Smyth's Sea – 580 × 550 km) straddling the eastern limb are visible. On the south-eastern limb, the interconnected mass of dusky patches that comprises sprawling Mare Australe

DAY 3

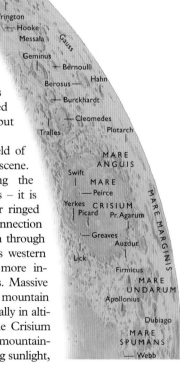

(Southern Sea), including the dark-floored crater Lyot (140 km), can be made out. East of Vendelinus is a large dark lava plain covering about the same area as Mare Marginis. The southern part of this large plain is marked by the flooded crater remnant Balmer (112 km), but the rest of it is currently unnamed.

With the entire Moon in the field of view, Mare Crisium dominates the scene. Mare Crisium is unique among the Moon's flooded multiringed basins – it is the only completely flooded major ringed basin that stands alone, without connection to any other large mare. Inspection through the telescope shows that Crisium's western mountain border is bulkier and more integrated than the eastern ramparts. Massive though they may be, the outer mountain ramparts do not seem to rise gradually in altitude towards the rim – instead, the Crisium basin seems to be sunk into a vast mountainous plateau. Bathed in early morning sunlight,

Mare Crisium's western wall appears sharply defined, bright and intricate, like a mass of pumice. From the middle of the western wall on the mare shore jut two opposing headlands (once known as Promontorium Lavinium and Promontorium Olivium), pointing at each other like two jagged mountain fingers.

It is easy to see the effects of the curvature of the Moon by noting how the illumination of Mare Crisium is greater in the east and gradually diminishes towards the terminator. Sometimes the mare is so poorly illuminated in the far west that the western mountain border appears to float like a bright curved line in black, empty space. As the Sun rises, wrinkle ridges can be traced on the middle and western parts of Mare Crisium's floor. The narrow Dorsum Termier winds for 90 km across the central southern floor; and the much broader Dorsum Oppel traces a 300 km path parallel to the western mare border, from the northern part of the mare, mingling with the eastern wall of Yerkes and then proceeding some distance south. Yerkes (36 km) is a flooded crater connected to a small crater 30 km to its north by a narrow mountain branch, but at morning illumination its eastern wall casts quite a solid shadow on to its floor. As the Sun rises, once-proud Yerkes soon vanishes into oblivion.

There are a number of other small craters within the western sector of Mare Crisium, including the diminutive Swift (11 km), Peirce (19 km) and Picard (23 km), and the bowl-shaped Greaves (14 km), plus the flooded Lick (31 km), which lies on the south-western shoreline.

Immediately north of Mare Crisium, the crater Cleomedes (126 km) has emerged into the morning sunshine, its relatively smooth grey floor punctuated by two craters, a small mountain and a forked rille, the latter requiring at least a 150 mm aperture to resolve. On Cleomedes' outer north-western ramparts, the complex crater Tralles (43 km) is interesting to study at a high power. Tralles was undoubtedly formed after Cleomedes, perhaps as a result of a multiple, near-simultaneous impact – at least three overlapping crater impressions can be traced in the one feature. To the north of Cleomedes lies another, much larger multiple impact crater, Burckhardt (57 km). With its main central crater and twin lobes, the crater resembles some early telescopic drawings of Saturn and its prominent 'ansae' in the era before Christiaan Huygens deduced the true nature of the rings. Another prominent crater in this region is Geminus (86 km), some 40 km north of Burckhardt. Geminus has a terraced internal wall and a mere sliver of a central mountain. A parallel pair of shallow valleys crosses the grey plains away from Geminus' southern rim to a distance of about 80 km, but these require a larger aperture to view at all well. Farther north of Geminus, the less prominent walled plain Messala (124 km) is on view, with its interesting undulating floor packed with craterlets and hills.

Between Mare Humboldtianum on the north-eastern limb and the three-day-old terminator, the large walled plain Endymion (125 km) has come into view. Several other walled plains – notably Plato and Archimedes – resemble Endymion with its near-circular outline and smooth dark floor, but none are nearly as large. At this early morning illumination, Endymion appears attached to a larger walled plain to its north-east, but this is a highly eroded feature and currently unnamed.

▲ *Vallis Rheita, a large valley made up of interconnected craters. This observational drawing was made on 23 March 2000 by Grahame Wheatley, using a 128 mm refractor.*

Mare Fecunditatis is bisected by the morning terminator, and close study will bring to light numerous wrinkle ridges: Dorsum Cushman, Dorsa Cato and Dorsum Cayeux in the north; the intricate Dorsa Geikie system (240 km long) farther south; and the Dorsa Mawson, 180 km long and in places more than 10 km wide. South of Mare Fecunditatis, the crater Snellius (83 km) is linked to one of the Moon's longest valleys – Vallis Snellius, a 500 km long chain of interconnected craters formed by the debris blasted out by the Nectaris impact. Under low magnification, the valley looks like a black line running from the terminator and ending at Snellius' western rim, but it actually cuts across Snellius' southern floor and continues towards the limb. Vallis Rheita lies about 250 km south of Snellius. It is more than 500 km long, exceeds 50 km wide in places, and cuts south-east across the southern highlands from the crater Rheita (66 km) to Reimarus (48 km). Like Vallis Snellius, it appears to be radial to Mare Nectaris and was probably formed as a result of secondary impacts from the asteroidal collision that created the Nectaris basin.

Day four

Mare Crisium and Mare Fecunditatis are both easy to see with the unaided eye. To the west, Mare Tranquillitatis (Sea of Tranquillity) is beginning to be exposed on the terminator, its dark eastern plains detectable with a keen naked eye. The earthlit hemisphere remains visible, though it is not as prominent as on day three, but binoculars will show the earthshine clearly enough.

DAY 4

Through binoculars, Langrenus, Vendelinus, Petavius and Furnerius in the south remain easily visible, Langrenus and Petavius being the most impressive of these craters. Both the Rheita and Snellius valleys can also be glimpsed in steadily held medium-power binoculars. The craters Cleomedes, Geminus, Messala and Endymion also remain on show. Langrenus is beginning to brighten relative to its surroundings, as Endymion appears to darken, and the near-limb eastern maria are also becoming more apparent. Even through binoculars, the exceedingly complex, crater-crowded nature of the southern uplands is becoming evident. Large craters, such as the conjoined twins Steinheil and Watt (both 66 km across) and the group of prominent large craters clustered around Hommel (125 km), are beginning to announce their presence towards the end of day four.

Binoculars reveal a bright spot just past the western mountain border of Mare Crisium. A higher-magnification view through a telescope will show a small, markedly polygonal crater – Proclus (28 km). As the Sun rises throughout day four, two particularly bright rays can be seen spreading away from Proclus, one to the north-west and the other to the south. Other, less prominent rays extend eastwards on to Mare Crisium itself. Between the rays and the eastern border of Mare Tranquillitatis is a grey diamond-shaped area called Palus Somni (Marsh of Sleep). Proclus, with its ray system, and Palus Somni will both become far more obvious as the morning shadows recede in the coming days.

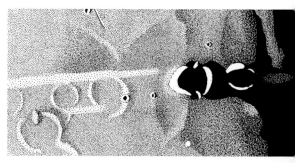

▶ The double crater Messier and Messier A in Mare Fecunditatis has a remarkable ray system. Observational drawing by Phil Morgan using a 305 mm Newtonian.

North-eastern Mare Tranquillitatis catches the rays of the early morning Sun. A northern projection of the mare has been named Sinus Amoris (Bay of Love), and north of this are Montes Taurus, a broad jumbled mountain plateau a couple of hundred kilometres in diameter with peaks that reach 3000 metres. A number of geologically interesting features occupy the north-eastern sector of Mare Tranquillitatis. Cauchy (12 km) is a simple bowl-shaped crater, but immediately to its north is Rima Cauchy, a wide rille 210 km in length that runs north-west/south-east across the plain in a slightly curved path. To the south lies Rupes Cauchy, an odd combination of fault-scarp (120 km long) and rille with an overall length of 180 km. Rupes Cauchy is also slightly curved, but in an opposite direction to Rima Cauchy – the pair have been referred to as 'hyperbolae'. The scarp face of Rupes Cauchy faces west and casts a thin shadow during the early morning illumination.

Farther south lie the large volcanic domes of Tau and Omega Cauchy, each with a base measuring some 15 km across and rising to several hundred metres. All these features can be discerned in a 60 mm telescope. Both domes are topped by a tiny summit crater – actually the remnants of the volcanic vent – the largest of which lies on Omega and is resolvable with a 150 mm aperture, giving the dome the appearance of a little pop rivet.

Mare Fecunditatis bathes in the morning Sun, but its border is not nearly as mountainous or as sharply defined as that of Mare Crisium. On Mare Fecunditatis' northern shore lies Taruntius (56 km), an exciting crater to view at high magnification, with its well-defined rim and hummocky, somewhat ill-defined concentric inner ring surrounding a small central peak. Taruntius' outer flanks are complex and will repay further detailed scrutiny at this low angle of illumination. Taruntius' rim remains visible

under a high Sun, and it is the centre of a fairly bright ray system which can be traced for more than 100 km. In the mare's north-western quadrant are two tiny craters, lying side by side – Messier and Messier A. Messier is elliptical, measuring 9 × 11 km and elongated east–west, while Messier A is an odd 13 × 11 km peanut-shaped double crater. A remarkable double ray stretches like two parallel searchlight beams more than 150 km westwards from Messier A to Fecunditatis' border. Messier's rays will become easier to see as the Sun rises. In the general vicinity of Messier, particularly to the south, a very low Sun will reveal a dozen or more ghost craters in Mare Fecunditatis. Some of these are independent, while others appear to be intimately connected with the more prominent wrinkle ridge systems. Many of these features are undoubtedly craters which have been completely buried by lava flows.

Mare Fecunditatis' south-western border is marked by the mountain spine of Montes Pyrenaeus (the lunar Pyrenees), and two prominent craters are located in the mare nearby – Gutenberg (74 km) and Goclenius (72 km). Both craters are irregular in outline, each having floors dotted with small peaks and crossed by rilles, part of the prominent Rimae Goclenius system that crosses north-west/south-east over the eastern plains of the mare. Rimae Goclenius are the first major marial rilles visible in the lunar month, but they require at least a 100 mm aperture to resolve adequately.

Towards the end of day four, the terminator exposes the huge crater Janssen (190 km) in the southern uplands, to the west of Vallis Rheita. Janssen has an epic feel about it – a crater that has been through the trauma of asteroidal bombardment many times yet retained its identity in the face of adversity. A couple of sizeable rilles cross Janssen's floor, easily visible in a 100 mm aperture. The largest is undoubtedly a graben produced by faulting, but the other appears decidedly sinuous and may be a lava-carved feature.

Day five

The broad crescent Moon of day five is a delight to view with the naked eye, as it can be seen high above the western horizon in a dark sky long after dusk, surrounded by the brighter stars and the occasional planet. The feeble glow of earthshine can just be made out, clinging to the narrowing lunar dark side like an after-image. Binoculars will show much more of Mare Tranquillitatis, although it will be another day until its western margins are revealed. The terminator has rolled back to unveil the whole of Mare Nectaris in the south, a near-circular sea about 210 km in diameter. North of Mare Tranquillitatis, the extreme eastern margin of Mare Serenitatis (Sea of Serenity) is on view, producing a 'dent' in the termina-

DAY 5

tor visible to the keen naked eye, and it is bordered in the east by the broad mountain range of the Montes Taurus. A gap in Mare Serenitatis' northern border joins it with the large dark plains of Lacus Somniorum (Lake of Dreams). Farther north, the eerie-looking Lacus Mortis (Lake of Death) joins with the south-eastern plains of Mare Frigoris (Sea of Cold).

Binoculars show that Endymion in the north-east has been joined by two large, deep-looking craters some distance to its south-west – the majestic duo of Atlas (87 km) and Hercules (69 km). Viewed at high magnification through a 100 mm aperture, a prominent system of rilles can be discerned on Atlas' floor, branching from the southern wall to perform a pincer movement around the crater's central mountain. Considerable slumping has occurred in Atlas' inner wall, producing terracing and quite a sharp rim. Immediately west of Atlas, the smaller Hercules makes a nice contrast, with its dark floor and large sharp floor crater, offset to the south. Lacus Mortis, 150 km in diameter, is located west of Hercules. It appears to be the flooded remnants of a large ancient crater that has experienced a considerable range of geological activity. The crater Bürg (40 km) is on the east of Lacus Mortis' floor, and is interesting in itself, with its deep floor, central mountain and relatively broad, terraced inner walls.

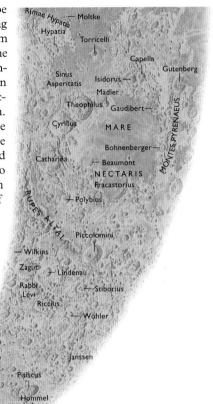

Wide ridges splay north and south of Bürg, and west of them can be seen Rimae Bürg, a system of at least seven rilles, two of which are easily visible in a 60 mm telescope. The overall appearance of Lacus Mortis is striking and rather creepy, giving the impression of an ancient cracked landscape that seems a little out of place on the Moon – perhaps more at home on Sun-baked Mercury.

A fairly bland and featureless scene is presented by the eastern part of Mare Frigoris, save for a few low wrinkle ridges. Several large, flooded walled plains are on view in the northern highlands, including Arnold (95 km) and Baillaud (90 km), but the area is far less dramatic in appearance than the southern uplands.

Beautiful Posidonius (95 km) marks the north-eastern border of Mare Serenitatis. Binoculars show that it has a complex floor and that

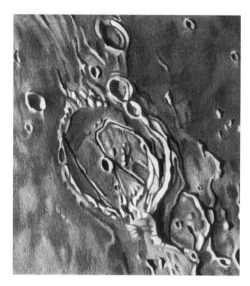

◀ Posidonius illuminated by an evening Sun. This observational drawing was made by Peter Grego on 29 July 2002, using a 150 mm achromat and based on a CCD image template.

it is joined to the south by the disintegrated crater Chacornac (51 km). Viewed up close through a 150 mm aperture, Posidonius' floor is revealed in all its splendour. An offset bowl-shaped crater, Posidonius A, is surrounded by a number of rilles, hills and ridges. A 100 mm telescope will hint at the detail, but the network of rilles requires a 150 mm instrument and fairly high magnification to be seen at its best. Posidonius' north-eastern wall is intruded upon by several craters that give the impression of a gradually diminishing chain. Chacornac also has a tiny offset floor crater and is striated with rilles. About 80 km east of Posidonius is the unusual Rima G. Bond, which cuts for 150 km across the border of Lacus Somniorum in two connected arcs, like a flattened W on its side.

Farther to the south, along Mare Serenitatis' eastern border, the flooded crater Le Monnier (61 km) makes an interesting semicircular indentation in the mountains. Under a low Sun the flooded western ramparts of this feature may be discerned as a low ridge which crosses the bay that Le Monnier makes. Nearby, the multiple linear rilles of Rimae Littrow, the longest of which measures some 170 km, cut across hill and dale around the tiny crater Clerke (7 km) and the disintegrated crater Littrow (31 km). Mare Serenitatis' south-eastern mountain border ends at the massive mountain block of Mons Argaeus, 50 km long and rising to 2000 metres above the marial plains to the west. As the Sun climbs higher through the fifth day, progressively more detail along the eastern plains of Mare Serenitatis comes into view, including a narrow braided wrinkle ridge 120 km in length, Dorsa Aldrovandi, which crosses the sea from Mons Argaeus to the promontory that marks the southern wall of Le Monnier. Just to the west of Mons Argaeus lies one of the Moon's least conspicuous craters – the ghost-like submerged crater Abetti, just 7 km in diameter, visible only at low angles of illumination.

The eastern sector of Mare Serenitatis is home to one of the most pleasing of the Moon's wrinkle ridges – Dorsa Smirnov (130 km long). It has its origin west of Posidonius in the north of the mare, and winds its way south to touch another prominent wrinkle system, Dorsa Lister (290 km long). Both ridges are in fact part of the same complex of ridges once known as the 'Serpentine Ridge' because of its snake-like appearance. Dorsa Lister curves westwards and then north-west over to the crater Bessel (16 km), and the ridges continue yet farther north to join with Dorsum Azara. There are more wrinkle ridges to be observed in this area under low illumination, and the scene is quite enthralling. In part, the ridges are attributable to the contraction of the marial surface after its formation, but some represent actual buried features, and appear to mark the position of an ancient inner basin ring that has been buried by lava flows.

A 200 km wide strait connects southern Mare Serenitatis with northern Mare Tranquillitatis, but three large parallel graben rilles, Rimae Plinius, spring eastwards from the sharp serrated cape of the Promontorium Archerusia to trace the southern outline of the Serenitatis basin. Several wrinkle ridges actually emanate from the southern part of Dorsa Lister and cross the marial divide, struggling to negotiate the 'cattle grid' of Rimae Plinius. High magnification with a 150 mm aperture will reveal that the rilles actually cut through the wrinkles in places. Abetti's identical twin, the ghost crater Brackett (9 km), sits atop the northernmost of the Rimae Plinius.

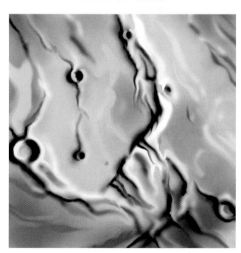

▶ Observational sketch of the magnificent wrinkle ridge Dorsa Smirnov in Mare Serenitatis, made by Peter Grego using a 200 mm SCT.

Towards the end of day five, the low morning Sun brings to light features of considerable selenological interest in western Mare Tranquillitatis. In the north, many wrinkle ridges radiate from a unique feature called Lamont, centred around a buried crater. Lamont is outlined solely by these ridges and is visible only when illuminated by a low Sun. The ridges surrounding Lamont were formed by marial compression around 3.5 billion years ago, when the fresh lava flows of Mare Tranquillitatis subsided. Arago (26 km) nestles among the ridges immediately north-west of Lamont. Just north of Arago is Arago Alpha, a lumpy volcanic dome some 15 km across, and to the west lies the slightly larger but equally lumpy dome Arago Beta. Neither dome has a discernible summit craterlet. South of Lamont, close to the southern shore of Mare Tranquillitatis, is Statio Tranquillitatis (Tranquillity Base), the site of Apollo 11's historic first moonwalk in July 1969. Three very tiny craters in the area have been named in honour of the Apollo 11 crew – from west to east, Aldrin (3.4 km), Collins (2.4 km) and Armstrong (4.6 km) – the only lunar craters ever to have been officially named after living people. The three craters can be glimpsed under a low Sun using a 150 mm aperture. Near the western shore of Mare Tranquillitatis lie Ritter and Sabine, two interesting craters with slight ridges on their floors. Larger instruments reveal fascinating detail when the area is lit by a low Sun, including a complex of parallel linear rilles near the south-western border of Mare Tranquillitatis, including Rimae Ritter (100 km long), proceeding north-west of Ritter, and Rimae Sosigenes (150 km long), skirting the south-western mare border.

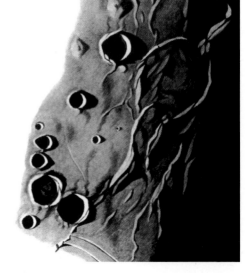

▶ *Lamont, an unusual feature of low relief in Mare Tranquillitatis. This drawing was made on 28 December 1999 by Grahame Wheatley, using a 240 mm reflector.*

▶ *The five-day-old Moon photographed on 26 December 1995 by Paul Stephens, using a 300 mm Newtonian.*

Moltke (6.5 km), a small bright-collared crater, can be seen south of Statio Tranquillitatis, beneath which extends the 180 km long Rimae Hypatia, named after the disintegrated, irregularly outlined crater Hypatia (41 × 28 km) some 60 km south of the rille. Hypatia lies on the north-western shore of Sinus Asperitatis (Bay of Asperity), a lava plain which displays a mass of wrinkles and ridges. Sinus Asperitatis serves as a bridge between southern Mare Tranquillitatis and northern Mare Nectaris. The pear-shaped crater Torricelli (23 km) lies within what appears to be the remnants of a larger flooded crater whose rim can be traced in a system of low hills from the north around to the south-west. Rough highlands dotted with small craters and small jumbled peaks lie to the east of Sinus Asperitatis, and this terrain is bisected by Rimae Gutenberg, visible through a 150 mm aperture. In the south, the craters Isidorus (42 km) and Capella (49 km) make a fine contrasting pair. Isidorus is fairly well-defined and has a flat grey floor pitted with a single crater-let, encircled by a low narrow rim, while Capella is more irregular in outline with a comparatively chaotic hilly floor. Capella is crossed by a roughly hewn crater chain, Vallis Capella (110 km long).

The whole of Mare Nectaris is on view during day five. With a surface area of 100,000 sq km, Nectaris is one of the smallest lunar maria, but while most nearside marial basins have been totally flooded with lava, Mare Nectaris actually resides in a larger basin around 700 km across. It is more like the partially flooded farside lunar basins such as Mare Orientale. Only the south-western rim of the wider Nectaris basin is clearly defined, marked by one of the Moon's most spectacular features – Rupes Altai (described below). Ghost craters occupy the northern part of Mare Nectaris' floor. Daguerre (46 km) is one of the Moon's loveliest ghost craters, an ancient buried crater whose southern wall has been utterly lost to sight beneath lava flows. Adjoining Daguerre in the north-west is a larger, unnamed ghost crater. South of the crater Gaudibert (33 km) is an intricate little grouping of ghost craters, including Gaudibert A and B. Several small wrinkle ridges proceed southwards along the eastern shore of Mare Nectaris, west of the crater Bohnenberger (33 km).

In the far south the mountain border sports the giant semicircular dent of Fracastorius (124 km), easily visible through binoculars. Fracastorius is an ancient crater filled with lava flows from Mare Nectaris, which have eroded much of its northern ramparts so that it has taken on the appearance of a great bay, with dimensions comparable to the terrestrial Gulf of Taranto at the foot of Italy. Numerous craters are embedded in Fracastorius' wall, the largest of which are in the west. A large telescope and good seeing will resolve the tiny cleft that winds east to west across the southern part of the crater floor, along with several tiny craterlets. North-west of Fracastorius lies Beaumont (53 km), another crater whose wall has been breached, admitting the lava flows of Mare Nectaris, although the breach is less than 10 km wide, and Beaumont has retained its overall outline.

Like a thick rope, a wide wrinkle ridge proceeds north from Beaumont across Mare Nectaris to the outer eastern flanks of Theophilus (100 km), which is now catching the rays of the morning Sun, along with Cyrillus (98 km) and Catharina (100 km). Sunrise over this linked trio of craters is spectacular – indeed, it is one of the major highlights of the lunation – and by the end of day five they are the most obvious near-terminator features visible in binoculars. Theophilus is an impressive impact crater, with broad terraced inner walls and a wide floor dominated by a sprawling central mountain complex comprising three main peaks – Theophilus Alpha in the south, Phi in the east and Psi in the west. The shadows cast by the eastern rim soon pull back to reveal these peaks, which rise to 1400 metres. A small crater resides inside Theophilus' western rim. The rim itself rises some 1200 metres above the surrounding landscape, and beyond it lie fields of prominent radial ridges and furrows, typical of so many large lunar impact craters.

Adjacent to the south-western wall of Theophilus lies Cyrillus, an older crater whose north-eastern flanks have been deformed by Theophilus. Cyrillus' rim is less sharply defined than Theophilus', and its walls and floor are more jumbled and chaotic in appearance. Like Theophilus, Cyrillus has a mountain complex on its floor with three main peaks – Cyrillus Alpha to the east, Eta to the west and

Delta to the south – and its inner western wall also has an embedded crater. A small rille cuts across the crater floor south of the central mountains. Catharina is connected to the southern wall of Cyrillus by 40 km of heavily striated landscape. Ancient and disintegrated, the northern part of the crater is occupied by Catharina P (55 km), whose presence opens up the main crater's northern ramparts to the terrain beyond.

Binoculars show the large prominent crater Piccolomini (88 km), over 100 km due south of Fracastorius. Piccolomini is slightly polygonal in outline, with broad inner terraces and a grey floor towered over by a very impressive central mountain massif rising to 2000 metres. Through binoculars a bright curving line is visible running from the western flanks of Piccolomini to the terminator. This is Rupes Altai, the nearside's largest fault feature, at almost 500 km in length. These mountains present a curving, scalloped scarp face of staggering proportions whose origin is linked to the stresses set up in the lunar crust by the asteroidal impact which gouged out the Nectaris basin 3.9 billion years ago. The inner part of the basin has subsided, exposing the scarp face along the line of a deep-seated fault. The young crescent Moon shows Rupes Altai as a bright winding line, in places up to 15 km wide.

In the far south, the moving terminator reveals more of the highly cratered southern uplands. By the end of day five the terminator has uncovered a number of interesting large craters in the mid-southern latitudes. South of the disintegrated Wilkins (57 km), the trio Zagut (84 km), Lindenau (53 km) and Rabbi Levi (81 km) huddle together. To their southeast lies Riccius (71 km), a crater that surely holds the record as the Moon's most heavily bombarded crater – it resembles a well-used artillery firing range, and the only remaining portion of the crater wall is in the west. In the far south lie the large walled plains Mutus (78 km) and Manzinus (98 km), and depending on how favourable the libration is, the near-limb craters Boguslawsky (97 km), Demonax (114 km), Schomberger (85 km) and Scott (108 km) may be visible, though identifiable only with the aid of a very detailed map or by an experienced eye.

▲ *Theophilus, Cyrillus and Catharina. The drawing was made on 9 April 2000 by Colin Ebdon, using a 250 mm reflector.*

Day six

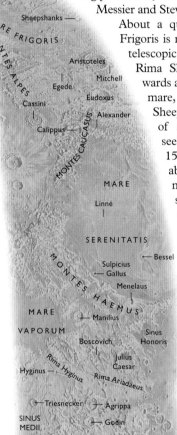

DAY 6

Through binoculars, tonal variations can now be discerned in Mare Serenitatis, its eastern and southern margins being noticeably darker and sharply delineated from the rest of the mare. At the beginning of day six the terminator lies across the western part of Mare Serenitatis, gradually sweeping westwards over the coming day to uncover the whole of the mare. In the north, the dusky tract of Mare Frigoris can be discerned with the naked eye.

Looking back over the features that are now experiencing the high mid-morning Sun in the east, several once prominent craters have now vanished without trace. Some bright ray systems in the east are becoming increasingly obvious, such as those around Proclus, Messier and Stevinus A.

About a quarter of the length of Mare Frigoris is now in sunlight. A good test of telescopic resolving power is the wire-thin Rima Sheepshanks, which cuts westwards across the northern plains of the mare, from just south of the crater Sheepshanks (25 km) for a distance of about 200 km. In excellent seeing conditions, a really good 150 mm instrument should be able to resolve this feature. Two major craters on Mare Frigoris' southern border really stand out at this angle of illumination – the close pairing of Aristoteles (87 km) and Eudoxus (67 km). Aristoteles is a most impressive formation – a clearly defined impact cater with strongly furrowed outer flanks and a broad, intricately terraced inner wall. Lava flooding has obliterated much of the crater's original central mountains – its floor is rather flat, with two very small peaks in the south. Adjoining the crater's eastern rim is the disintegrated crater Mitchell (30 km), possibly more ancient than Aristoteles. Some 70 km to the south, the crater Eudoxus nestles comfortably within the hilly terrain. Eudoxus also has a wide terraced inner wall and a few minor peaks on its floor, but its outer flanks are not as obviously sculpted as its neighbour's. West of Eudoxus are the eastern foothills of the vast mountain range of Montes Alpes (the lunar Alps), which are beginning to display some of their grandeur. An interesting crater named Egede can be seen 100 km north-west of Eudoxus – a low-walled dark plain that looks out of place in this jumbled upland environment. To the south is Alexander, one of the least crater-like of the Moon's craters, a small dark plain some 82 km across; sharp irregular mountains, Montes Caucasus (Caucasus Mountains), mark its western boundary. These mountains mark the boundary between north-western Mare Serenitatis and Mare Imbrium (Sea of Rains), and in the middle of day six the highest of the range's peaks (some of which rise to 6000 metres) can be seen jutting out of the blackness past the terminator.

Mare Serenitatis is gradually exposed during the sixth day, and its odd shape becomes apparent. The western shoreline is only very slightly curved, but veers off at a sharp angle in the north, giving it a square shoulder. Here can be seen the polygonal crater Calippus (33 km) and the tiny Rima Calippus (40 km long) some distance to its south-east, near the mare shoreline. In Mare Serenitatis, just east of the most southerly Caucasus peaks, lies a large dome with a base about 30 km in diameter. Several very small peaks poke through the top of the dome – an unusual topography rarely seen elsewhere on the Moon. It has no official name, but many lunar observers refer to it as the 'Valentine' dome, presumably because some observers imagine that it appears slightly heart-shaped.

Mid-western Mare Serenitatis is home to the diminutive crater Linné (2.4 km), which is surrounded by a bright circular spot 10 km in diameter that is easily visible in binoculars. A 150 mm aperture is needed to reveal the crater itself, and this is possible during day six when its interior is filled with early-morning shadow. Linné occupies an infamous place in lunar observational history. It was once cited as evidence that the lunar surface can undergo permanent change, for several highly respected lunar observers claimed that Linné had once been a much larger crater. Sadly, there's no real evidence to support these claims, and the Linné affair seems to have been a simple case of misinterpretation of historical observations.

Under early-morning illumination, the striated mountains of Montes Haemus are a magnificent sight. The range stretches in a nearly straight line along Mare Serenitatis' south-western margin for around 400 km, its peaks separated here and there by wide linear valleys and dark elongated lakes such as Lacus Odii (Lake of Hatred), Lacus Doloris (Lake of Sorrow) and Lacus Gaudii (Lake of Joy). Near the shoreline lie the bright bowl-shaped crater Sulpicius Gallus (12 km) and a rille system connected to it; the system runs 90 km north-west and branches out like a witch's broom brushing the coastline.

South of Montes Haemus lies Manilius (39 km), a prominent crater somewhat polygonal in outline with internal terracing and a small central peak system. Under the early morning sunlight, the binocular observer will notice a broad dark plain to its west, and the unsuspecting viewer might imagine that another great sea is about to emerge into daylight. In fact, this plain consists of the small Mare Vaporum (Sea of Vapours) to the west and the equally small Sinus Medii (Central Bay) to its south, both of which have a surface area of a little more than 50,000 sq km. Mare Vaporum is a rather bland place, its plains containing virtually nothing of observational interest save a few very minor wrinkle ridges. Sinus Medii, however, contains a veritable treasure trove of lunar gems (see 'Day seven').

The whole of Mare Tranquillitatis is now visible. Through a telescope, plenty of detail is discernible to its west. The western mountain flanks of Tranquillitatis possess a striated appearance – these are deep-seated linear forms that radiate from Mare Imbrium. They represent the 'Imbrium sculpture' which was etched deep into the lunar landscape during the formation of the giant Imbrium basin about 3.8 billion years ago. The entire area, from Montes Haemus on the south-western border of Mare Serenitatis to the centre of the lunar disk, looks like a landscape that has undergone glacial erosion of ice-age proportions. Sinus Honoris (Bay of Honour) cuts into the western mountain border of Mare Tranquillitatis and flows along the sculpted lines, leaving several isolated linear hills jutting above the mare. A number of broad, dark, lava-filled linear valleys lie to the west, including Lacus Lenitatis

▶ *The six-day-old Moon photographed on 8 May 2003 by Peter Grego, using a 150 mm achromat and a Ricoh RDC-5000 digicam.*

(Lake of Tenderness, 80 km long), Boscovich (46 km across) and Julius Caesar (90 km). Other features in the area appear to buck the general north-west/south-east trend of the landscape. The prominent 220 km long rille Rima Ariadaeus runs westwards through the hills south of Julius Caesar and into the plains north of the crater Agrippa (46 km).

Southern highland features brought into view along the terminator during day six include the neatly formed walled plain Abulfeda (62 km), which makes a fascinating contrast with its neighbour to the north-east, Descartes (48 km), a far more ancient and exceptionally disintegrated crater. About 200 km farther south lies Sacrobosco (98 km), a walled plain with a complicated inner wall and three sharp-rimmed craters on its floor. Nearby, an interlocking trio of craters, Azophi (48 km), Abenezra (42 km) and Abenezra C (48 km, but considerably overlapped in the east by Abenezra), is fascinating to view at a high magnification. Abenezra has an unusual swirling pattern of ridges on its floor, looking as though it has been disrupted by some mighty lunar vortex. Gemma Frisius (70 km) sports four large craters on its northern margin, and looks like a huge paw-print, with Goodacre (46 km) as the big toe. Farther south, the walled plain Maurolycus (114 km) is a magnificent example of a very large crater that has been planted almost – but not quite – over an equally large crater, the remnants of which can be observed to the south. The crater Barocius (82 km) also overlaps this ancient crater, and snuggles up to Maurolycus' south-eastern wall.

Day seven

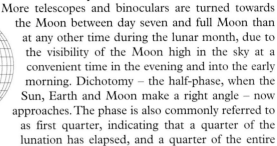

DAY 7

More telescopes and binoculars are turned towards the Moon between day seven and full Moon than at any other time during the lunar month, due to the visibility of the Moon high in the sky at a convenient time in the evening and into the early morning. Dichotomy – the half-phase, when the Sun, Earth and Moon make a right angle – now approaches. The phase is also commonly referred to as first quarter, indicating that a quarter of the lunation has elapsed, and a quarter of the entire lunar surface is now on view.

Several large walled plains are visible in the far north. A favourable libration will bring Nansen (122 km) into view near the far northern limb. Nansen, like a handful of other lunar craters near the Moon's north and south poles, has been named in honour of a famous terrestrial polar explorer. In the far north lie the craters Peary (74 km), Byrd (94 km) and Gioja (42 km), which at a favourable libration can be discerned as extremely fore-shortened ellipses. The north pole itself actually lies within a small crater on Peary's northern rim – a good target to aim for during a favourable libration. To the south of Gioja lie the linked duo Main (46 km) and Challis (56 km), while to the south-east is Scoresby (56 km), all of which are reasonably easy to locate and identify. Scoresby is one of the few clear examples of a large lunar crater that overlaps a smaller one – this ancient and eroded unnamed crater measures some 30 km across, and its north-eastern sector is completely overlain by Scoresby. The fascinating plain of Meton (122 km), with its smooth grey floor, lies south-east

of Scoresby. Its outline is deformed by large crater bays, giving it a highly unusual bulbous appearance.

The walled plains in this area all have a muted appearance – their floors, and the surrounding landscape, have been flooded with lava. Adjoining Meton in the south-west is Barrow (93 km), which when viewed from above is one of the squarest-looking of all the Moon's walled plains. Spanning the ground between Barrow and the northern shore of Mare Frigoris is the irregular-shaped plain W. Bond (158 km). Viewed under early-morning illumination, it looks impressively solid, and rather more substantial than it actually is, but as the Sun climbs higher in the Moon's sky the hilly floor and ragged walls of W. Bond are revealed in all their disintegrated glory. With a 200 mm aperture, a narrow linear rille can be traced for 60 km across the central eastern part of the floor. W. Bond's south-eastern periphery is marked by a 10 km wide flooded valley. The polygonal crater Timaeus (33 km) perches on W. Bond's south-western wall and surveys the plains of Mare Frigoris across to the Montes Alpes, 175 km to the south. Archytas (32 km) and Protagoras (22 km) are two sharp-rimmed but somewhat misshapen craters whose dark shadow-filled eyes keep watch over the northern approaches of Mare Frigoris.

Through binoculars, the broad mountain wedge of Montes Caucasus projecting south from Aristoteles and Eudoxus remains prominent during day seven – the range may even be discerned with a very acute unaided eye. A medium-magnification view of the series of lengthy, pointed black shadows being cast westwards by the Caucasus is simply superb. As the mountains' shadows recede, and the terminator rolls westwards, a new mountain range is revealed – Montes Alpes. The alpine lowlands gradually rise towards the west, reaching heights of 2400 metres. Here we find one of the Moon's most impressive showpiece features – Vallis Alpes (Alpine Valley), a unique feature that slices cleanly through 130 km of the lunar Alps from Mare Imbrium in the south to Mare Frigoris in the north.

Vallis Alpes is the best example of a lunar rift valley; it is similar in origin to many of the Moon's graben rilles, but is so wide (up to 20 km in places) that it is never referred to as a rille. Lavas flowing from the maria on either side of Montes Alpes have buried the original valley floor. A V-shaped bay in Mare Imbrium leads to a narrow pass in the Alps, marking where the valley proper begins. Vallis Alpes then opens out into a broad quadrilateral plain about 15 km wide, squeezing through another pass before it broadens again to cut north-eastwards across the Alps, widening to 20 km before gradually tapering nearly to a point on the shore of Mare Frigoris. A smaller rille has been carved down the middle of the resurfaced valley floor by the erosive action of very fluid lava flows, and southern parts of this narrow medial rille can be discerned through a 200 mm aperture when the Sun has risen high enough to illuminate the valley floor. The valley's walls have a sharp scalloped appearance along their entire length – nowhere are they straight, like those of a linear graben rille. A small cleft at right angles to Vallis Alpes (cutting across the valley floor itself) has been depicted by numerous telescopic observers, but in reality there is no such clear-cut feature, though a very rough ditch combined with the chance alignment of a couple of craterlets and peaks is just about traceable in Montes Alpes at a low angle of illumination.

To the north of Vallis Alpes is a particularly large, prominent mountain block (an unnamed feature) that rises a couple of thousand metres above the surrounding uplands. Clustered south of the valley are at least a dozen large bright Alpine peaks which catch the morning sunlight – they include Mons Blanc (3600 metres high) and the lofty capes of Promontorium Deville and Promontorium Agassiz. Mare Imbrium's far eastern plains are beginning to enjoy the morning light. Near the north-eastern Imbrium shore, between the Alps and the Caucasus mountains, an odd-looking crater comes into view. Cassini (57 km) is slightly polygonal in shape, and has a sharp rim. Its outer flanks consist of a broad collar up to 20 km wide, consisting of low, smooth hills, a kind of feature rarely found on the Moon, resembling many of the 'splatter' craters of Mars. Though Cassini has low walls, it casts a broad shadow with several pointed fingers westwards on to the mare, towards Mons Piton (whose summit is catching the sunlight beyond the terminator) at low illumination. Cassini's floor appears to be at the same level as the surrounding mare, and contains two size-able, sharp-rimmed craters. The largest of these, Cassini A (17 km), has a complex floor, and the illusion of a smaller crater within led to it once being nicknamed the 'Washbowl'. A winding double cleft some 20 km in length lies to the east of Cassini A, but requires at least a 200 mm aperture to resolve adequately.

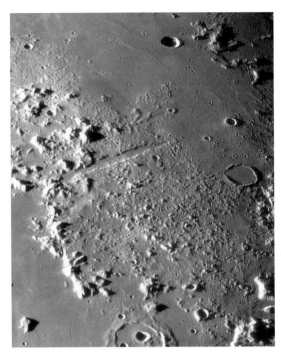

▶ *Vallis Alpes slices cleanly through a major mountain range. The image was obtained on 19 October 2000 by Mike Brown, using a 370 mm reflector and a HX516 CCD camera.*

South-east of Cassini is the neat little polygon of Theaetetus (25 km). This crater is surrounded by several low hills and ridges, those south-west of it being part of the radial impact structure surrounding Aristillus (55 km), which, along with Autolycus (39 km) to its south, is emerging into the morning sunlight (see 'Day eight' for a detailed description of the pair). Mare Imbrium's south-eastern border is marked by the impressive range of the Montes Apenninus (the lunar Apennines), the northernmost peaks of which are gleaming brightly on the terminator. It will be another day before the true majestic extent of the Montes Apenninus becomes apparent; at this low angle of illumination it is clear that the mountains are striated radially to Mare Imbrium, and this linear sculpture can be traced in the local terrain for more than 500 km to the south-east, with the exception of the flat plains of Mare Vaporum.

Curious dark lumpy terrain proceeds south from the border of Mare Vaporum, and here can be observed a feature which some observers have unofficially designated the Spiral Mountain or Mount Schnecken-berg. This formation, lying about 30 km north of Hyginus, is not at all obvious. It resembles an ancient, eroded crater about 30 km across with breaches in its northern and southern ramparts, with a domed floor and a small central depression. At high magnifications, the eye wanders to the south to linger over Hyginus (11 km), a sharply defined, pear-shaped crater at the centre of a superb valley – Rima Hyginus – a graben rille 220 km long intermittently punctuated with craters. Rima Hyginus can be made out with a 60 mm aperture, but a line of more than half a dozen linked craterlets along the rille to the west of Hyginus requires at least a 150 mm aperture to resolve.

About 100 km of Sinus Medii's plains separates Hyginus from Triesnecker (26 km), a lovely, deep polygonal crater with a sharp rim, terraced internal walls and a small group of central peaks. But the area between Hyginus and Triesnecker is by no means plain. Here lies a fabulously intricate network of fault valleys: Rimae Triesnecker, the finest rille system on the Moon, consists of around a dozen interlinked, criss-crossing valleys in an area about 10,000 sq km in extent, stretching from the northern border of Sinus Medii, south to the eroded crater Rhaeticus (45 km). A 100 mm telescope shows the main rilles in the network, but nothing can beat a high-magnification view of sunrise over the region through a large telescope under excellent seeing conditions.

Several wide bays lie along the southern border of Sinus Medii, including Flammarion (75 km), Réaumur (53 km) and an unnamed bay 80 km across in which the bowl-shaped Seeliger (9 km) resides. Nearby, the pathetic, barely visible remnants of a deeply buried 43 km crater have been named Oppolzer, and Rima Oppolzer cuts across its southern margin to the south-eastern border of Sinus Medii.

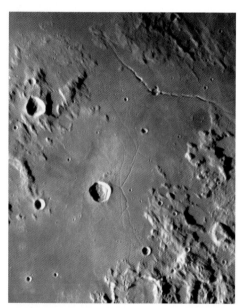

◀ Triesnecker and its extensive rille system (centre), with the Hyginus rille at top. The image was obtained on 19 October 2000 by Mike Brown, using a 370 mm reflector and a HX516 CCD camera.

Morning sunlight now illuminates the first of the big walled plains near the centre of the Moon's disk. Hipparchus (150 km), a disintegrated plain, lies immediately to the south-east of Sinus Medii. At sunrise it is a most impressive feature, its eastern wall appearing sharper and more solid-looking than its west wall. Several wide valleys slice from north-west to south-east through Hipparchus' eastern wall – clear examples of the radial Imbrium sculpture described earlier. When the western wall has emerged into the morning sunlight, the shadow cast by its eastern rim will have rolled back sufficiently to reveal the complex nature of the floor. Closer examination of what appears to be a group of peaks in the south shows them to be part of the southern ramparts of a heavily eroded, partly buried crater, Hipparchus X. To the east are several parallel rows of hills, from which a narrow rille runs into the north-western sector of Hipparchus' floor. There lies the remnants of a buried crater, visible only at a very low angle of illumination. The sharp polygonal crater Horrocks (30 km) dominates Hipparchus' northern floor. A number of large, bright circular bowl-shaped craters are dotted about the region, including Pickering (15 km) to the north-east, Hipparchus G on the eastern wall, and Hipparchus C and L in the highlands to the south-east. Adjoining Hipparchus' southern wall is Halley (36 km), a little walled plain with a flat dark floor, and to its east lies Hind (29 km), a more eroded crater.

Binoculars clearly show both Hipparchus and Albategnius on the morning terminator. Albategnius (136 km) is a grand walled plain of dramatic appearance, one of the most pleasing to scrutinize at medium to high powers. Much of Albategnius' floor is deep in the shadow cast by the crater's eastern rim, but the shadow is punctured by the gleaming sunlit heights of the large mountain Albategnius Alpha, located somewhat off-centre to the west. Albategnius Alpha is an interesting central peak, 20×10 km and oriented north–south, its summit reaching a height of 1500 metres. A 200 mm instrument will resolve the mountain's tiny summit craterlet. Two smaller knobbly lobes emanate from the northern part of the central mountain, one going north-east, the other south-west, both about 10 km long.

Albategnius' south-western wall is completely overlain by the crater Klein (44 km), whose eastern wall looks like a bright dented sickle at low illumination, when the northern part of the wall is in shadow. As the Sun rises, Klein's central peak becomes visible, and then its floor, which is flooded to the same level as the interior of Albategnius. The shadows on Albategnius' floor draw back to reveal a relatively smooth dark surface, but through a 200 mm telescope the area east of the central peak is seen to be a little dented here and there. A noticeably elongated crater, Albategnius B, adjoins the inner northern wall, where there is a mass of smaller craters. Albategnius' inner wall, from the north around to Klein, is extremely complex, crossed by a chain of a dozen or so 5–10 km diameter craters that reach 80 km north from Albategnius KA. The low illumination will also show up a deep groove in the north-eastern outer flanks of Albategnius; this valley, 8 km wide in places, runs south from the southern rim of Halley for about 85 km.

Parrot (70 km), an exceedingly ancient battered crater, connects to Albategnius' southern border and is the origin of another prominent valley, which runs south for some 60 km. To the east lies a lovely crater chain of gradually diminishing size, consisting of eroded, square-shouldered Airy (37 km), north through the sharp-cut Argelander (34 km), to Vogel (27 km) and Vogel B (20 km). The plain east of Vogel is home to the highly irregular crater Burnham (25 km), whose southern wall is breached by a wide valley 15 km long. When Airy lies on the morning terminator, the low angle of illumination brings into view a large walled plain – designated Airy M, 100 km wide – at the centre of which lies a smaller central ghost crater, Airy G. A similar sized, though far more prominent but unnamed walled plain lies farther south along the terminator. Playfair G measures about 110 km across, its floor dark and hilly, with a complex inner western wall and an eastern wall overlain by Playfair (48 km). To the south lies the neatly formed walled plain of Apianus (63 km), whose smooth dark floor appears noticeably convex under early-morning illumination, the western part of the floor appearing noticeably shaded.

During day seven the long string of craters from La Caille (68 km) in the north, through Blanchinus (58 km) and Werner (70 km) to Aliacensis (80 km), comes into view, easily identifiable in binoculars. The craters are quite different from one another, and make an interesting study in comparison. La Caille is a neat walled plain with somewhat eroded ramparts surrounding a smooth grey floor. Its northern and north-eastern walls are fringed with large craters, including Delaunay (46 km), which with its associated craters resembles the interior of a well-scooped tub of ice cream. Blanchinus is considerably more eroded, with a low, irregular wall and a patchy, hummocky interior. Werner is the cleanest-cut member of the chain, a near-circular formation with a sharp rim, broad, well-terraced interior walls, and a floor containing a group of small mountain peaks. Aliacensis' far northern and southern walls are partially disintegrated, but the other parts of the wall are well defined, and its interior terraces are wide and complex. Its slightly convex floor has a small peak, offset to the north. A little to the east lies Poisson (42 km), a highly deformed crater which together with Poisson T resembles a broad bean.

As the large walled plain Walter (135 km) emerges into the sunlight, bright spots poke out of the interior shadow into the sunlight. By the end of day seven they emerge as a compact cluster of craters and peaks, occupying Walter's north-eastern floor; the southernmost peak casts a broad wedge of shadow long into the lunar morning. Fairly smooth lava fill takes up the remainder of the interior plain, spotted with a few small craters, including the flooded Walter E near the base of the interior western wall. A large double crater, Walter L and K, dents Walter's eastern rim, and south of them lies Nonius (70 km), a disintegrated crater with a battered wall and a wide raised platform of a northern interior wall.

About 100 km farther south, Stöfler (126 km) and Faraday (70 km) make a nice connected pair. The well-preserved Stöfler has a reasonably sharp rim and clearly defined interior terracing. Two large crater bowls sit on its wall, Stöfler K in the north-west and Stöfler F in the south, both 20 km across. Through a 60 mm telescope the floor appears smooth and level, but a 150 mm aperture will resolve a trio of tiny craterlets in the north. Faraday overlies the entire south-eastern quadrant of Stöfler, and particularly noteworthy is a prominent jagged ridge which curves from Faraday's western wall to almost touch Stöfler's eastern wall. Though Faraday is younger than Stöfler, it appears more eroded, its south-western wall covered by an interesting group of large, deformed craters. Here, Faraday C (30 km) will repay close attention, a distinctly hexagonal crater with a cluster of little hills and domes on its floor.

Farther south, Licetus (75 km), Cuvier (75 km) and Heraclitus (90 km) make an intriguing group, Heraclitus forming the link between the trio. Heraclitus' southern wall is wholly occupied by Heraclitus D, whose floor level lies beneath the crater it has intruded upon, and its lack of a raised rim makes it appear to have simply sunk into Heraclitus. A prominent mountain spine runs from Heraclitus D's northern rim to completely bisect Heraclitus – one could say that Heraclitus is one of the few lunar craters with 'backbone'! South of Heraclitus, a chain of a dozen or more sizeable craters progresses eastwards for about 400 km, two prominent links in the chain being Lilius (61 km) and Jacobi (68 km). Lilius is a sharp-rimmed polygon with wide terraced internal walls and a very substantial single pyramid-like central mountain surrounded by a flat grey floor. Viewed from above, its near-neighbour Jacobi is distinctly rhombus-shaped, and its floor is crossed by a diagonal line of six 5–10 km wide craters. South-west of Lilius, on the meridian, is a curious flooded triple crater with an arc of peaks on its floor. Other craters worthy of note in the far southern highlands are the clear-cut Zach (71 km), the eroded plain of Curtius (95 km) and the square-shouldered Simpelius (70 km). Malapert (69 km) is the largest named crater near the lunar south pole – closer even than Amundsen – and traces of its rim and shadow-filled interior can be discerned at a favourable libration given a really detailed map and lots of patience.

Day eight

The eight-day-old Moon is a wonder to behold, the terminator dominated by many grand features. Binoculars show the magnificent mountain arc along the eastern border of Mare Imbrium, taking in Montes Alpes in the north, through Montes Caucasus along to Montes Apenninus in the south-east. A keen naked eye may be able to discern this curve as a wide dent in the terminator, including the southern heights of the Apennines where they peer into the darkness past

DAY 8

the morning terminator. A trio of large craters and a mountainous plateau dominate the eastern Imbrium plains, and the terminator to the south is packed with craters, with some exceptionally large and prominent ones on display.

In the far north the large plain Goldschmidt (125 km) is on view, along with its smaller, sharper companion Anaxagoras (51 km), on Goldschmidt's western wall. Anaxagoras is the centre of a bright ray system, and though the true extent of these rays becomes evident only under a high Sun, traces of them can be discerned across the floor of Meton and W. Bond even at this low angle of illumination. South of Anaxagoras, a very ancient and heavily eroded plain (unnamed) is visible, but it soon disappears as the Sun climbs higher. South of this feature lies the highly disintegrated plain of Birmingham (98 km), with its ill-defined walls and rubble-strewn floor – some of the roughest-looking terrain on any of the Moon's walled plains. Birmingham contrasts with nearby Epigenes (55 km) to its north-east, a sharp-rimmed crater with a

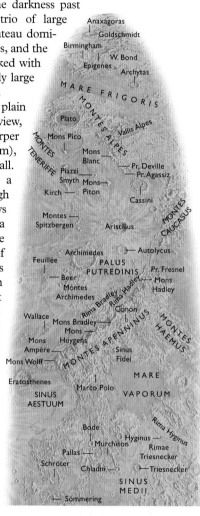

smooth dark floor, circular for most of its outline save for the south-east, where it forms a sharp elbow. A broad crater chain valley adjoins Epigenes to its south and runs 100 km to the north-east, forming the north-western wall of the large plain W. Bond. South of Birmingham, a wide mountainous headland presses down into northern Mare Frigoris, and a number of wrinkle ridges appear to converge on this promontory from both the east and the west. The topography of Mare Frigoris in this region appears to undulate, giving rise to several large dome-like elevations that can be discerned only at a low angle of illumination.

Vallis Alpes is clearly visible now, and Montes Alpes to its west contain some impressive large mountain blocks, the summits of which shine brightly in the morning Sun. Early in day eight, the Alpine terminator reveals a large dent. As the Sun climbs higher, the dent becomes a large, well-defined crater – the magnificent Plato (101 km), one of the loveliest of the Moon's walled plains. Almost circular in outline (but foreshortened somewhat by its northern location), Plato sinks into the Alps, its floor some 2000 metres beneath the level of its sharp rim. The rim itself does not rise very high above the surrounding mountainous landscape, perhaps up to 500 metres in places. A couple of indentations in Plato's western wall mark where the original rim has broken away and carried with it some very substantial pieces of crust towards the floor. The southern rim lies about 20 km from Mare Imbrium, and outside the south-eastern rim a prominent valley cuts through the hills down towards the mare shore. Plato's floor appears perfectly flat and featureless through a 60 mm telescope, with nothing of interest except for some subtle tonal variations. A 100 mm aperture will resolve several small craterlets dotted about the plain, visible as tiny bright dots under a high angle of illumination. The largest of Plato's craterlets lies just east of centre, with two close ones 25 km to its north-

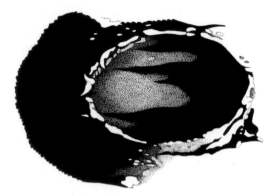

◀ *Morning shadows race across Plato's dark, flat floor in this observational drawing made by Nigel Longshaw on 18 June 2002, using a 125 mm Maksutov.*

west and another 25 km to its south-west. Under this early-morning illumination, the shadows cast by Plato's eastern rim on to its flat floor are fascinating to observe – a row of a dozen or more sharp black points, several of them longer than the others, whose eastward recession can almost be detected in real time as the Sun climbs higher.

A 150 mm telescope will resolve several rilles in the mountains between the eastern rim of Plato and the Vallis Alpes. The most promi-nent are Rimae Plato, which wind eastwards from near the craterlet Plato G for about 70 km. North of the crater Plato is a seldom mentioned valley of almost the same dimensions as Vallis Alpes; it has retained its original floor but is much more poorly defined. Sadly, this 'Vallis Alpes Minor' is oriented nearly east–west, and may only be seen when close to the terminator.

Prominent wrinkle ridges and mountain peaks have come into view on the Imbrium plain to the south of Plato, including a feature unoffi-cially known as Ancient Newton which gives the impression of a crater that has been buried deep beneath the Imbrium lava flows. Ancient Newton's western 'wall' is made up of the easternmost peaks of Montes Teneriffe (Tenerife Mountains) and Mons Pico to the south. The largest eastern peak in Montes Teneriffe always reminds me of a little trilobite (facing east), complete with a bifurcated tail. To the west, the main part of the group is a wishbone-shaped island, with a large mountain clump lying to its north-west. The entire combined length of the range is 110 km, and its highest peaks reach 1800 metres.

Mons Pico, though not officially classed as part of Montes Teneriffe, is named after Pico de Teide, a mountain on the terrestrial Tenerife islands, which, at 3715 metres, is the world's third highest active volcano. The lunar Mons Pico is a little smaller – 2400 metres high, with a base measuring 15 × 25 km. Pico throws a long, double-tipped shadow westwards on to the Imbrium plain, just south of Montes Teneriffe. Mons Pico Beta lies 60 km south, and it too is capable of throwing a lengthy shadow on to the Imbrium plain in the early morning or late evening. A broad wrinkle ridge with a prominent crest runs south from Mons Pico Beta, and another links the buried eastern 'wall' of Ancient Newton to the prominent bowl-shaped crater Piazzi Smyth (13 km). The shadows cast by the southern Alps, including Mons Blanc, Promontorium Deville and Promontorium Agassiz, are now receding from the eastern Imbrium plains. To their southwest, and just 40 km from Piazzi Smyth, lies the isolated mountain peak of Mons Piton (another lunar peak named after a mountain in the terrestrial Tenerife Islands). Like Mons Pico, Piton has a more southerly companion – Mons Piton Gamma, a slender twig of an island 15 km long.

Eastern Mare Imbrium is host to a lovely quartet of features – the large walled plain Archimedes (83 km), the craters Aristillus (55 km) and Autolycus (39 km), and Montes Spitzbergen (Spitzbergen Mountains) – all of which can be followed throughout the lunar day,

their outlines being easy telescopic objects under a high Sun. Aristillus and Autolycus emerge into the morning sunlight first. Both craters are surrounded by dramatic radial impact sculpture, much more prominent around Aristillus, which has a 30 km wide flange of sharp radial ridges interlaced with concentric ridges, and several low radial ridges reaching out as far as 100 km. Immediately north of Aristillus and almost obscured by the radial ridges is a 30 km diameter ghost ring, most probably a crater which was almost completely wiped out by the Imbrium lava flows long before Aristillus was formed. Aristillus is a classic impact crater, with wide inner walls displaying well-defined terracing, and a flattish floor dominated by a large central group of mountains. Rising to heights of 900 metres, the central peaks catch the morning sunlight before any of the floor is illuminated. About 40 km to the south, the sharp-rimmed Autolycus is slightly irregular in outline, and its floor is decidedly bumpy, devoid of any significant elevations. The crater's north-eastern rim hosts a small 3 km diameter crater (a test object for a 100 mm telescope) which appears initially as a tiny black notch.

Several hours after Aristillus and Autolycus have emerged from the terminator, the eastern flanks of Archimedes become illuminated

▲ *Morning over eastern Mare Imbrium. This digicam image was obtained on 1 April 2001 by Doug Daniels, using a 150 mm achromat.*

along with the bright peaks of Montes Spitzbergen. Circular in outline with a flat dark floor, Archimedes is superficially similar to Plato (some 500 km to the north), but its outer ramparts are more clearly delineated as they rise from the level mare, and some of its original radial impact texture is still visible. Archimedes' floor is striated with several light-coloured rays emanating from Autolycus to its east. Several small craterlets are dotted about the floor, notably to the west, but they require a 150 mm aperture to resolve. Apart from these craterlets the floor appears perfectly flat, even at very low angles of illumination and when seen through a larger telescope under high magnification.

Montes Spitzbergen, 60 km from Archimedes' northern rim, is a lovely compact mountain range. Stretching 60 km from north to south, it consists of a string of more than a dozen individual island peaks which tower up to 1500 metres above the flat Imbrium plain. The range casts a broad black shadow westwards into the morning terminator; several hours later, as the terminator moves on, the shadow detaches to become a series of pointed fingers, which mingle with the shadows cast by the broad buckles of wrinkle ridges to the west. At low angles of illumination, viewed at a medium magnification to take in the surrounding area, it becomes obvious that Montes Spitzbergen are not all individual peaks, but form part of a ridge complex that marks an ancient, largely buried inner ring of Mare Imbrium.

Palus Putredinis (Marsh of Decay), south-east of Archimedes, is a dark lava plain with an area equivalent to the island of Cyprus; it is notable for the ghost crater Spurr (13 km) some 30 km from Aristarchus' outer flanks, and a wrinkle ridge to its west. A group of rilles, Rimae Fresnel, run more than 50 km north from the margin of Palus Putredinis to Promontorium Fresnel. At the extreme eastern border of Palus Putredinis, near the foot of the towering Mons Hadley in the lunar Apennines, lies Rima Hadley, a sinuous rille 80 km long. South of Archimedes spreads a moth-eaten blanket of mountain peaks – the sprawling plateau of Montes Archimedes, covering around 45,000 sq km and containing some individual peaks that exceed heights of 2000 metres. The lowlands to the east of Montes Archimedes contain several rilles aligned north-west/southeast: Rimae Archimedes (difficult to discern, even with a 200 mm aperture) nearly intersect with the broader, more clearly defined Rima Bradley. Resolvable with a 100 mm aperture, Rima Bradley is a graben rille which undulates north of the Apennine foothills for 130 km, following the mare margin. Bancroft (13 km), a sharp crater bowl 2500 metres deep, indents the north-western margin of Montes Archimedes, surrounded by several shallow valleys. Some 50 km to its west are the twin crater bowls of Feuillée and Beer (both 10 km) from which runs a lovely little crater chain visible in a 200 mm aperture, along with a neat little dome to the south.

▶ *The eight-day-old Moon photographed on 21 February 2002 by Peter Grego, using a 100 mm Maksutov and a Ricoh RDC-5000 digicam.*

The region to the south of Montes Archimedes is distinctly lumpy, with several larger but less distinct dome-like features. Between the southern edge of Montes Archimedes and Montes Apenninus, the Imbrium plain displays some very interesting features. Two domes, connected by a thin ridge 40 km long, lie just north of the tiny crater Huxley (4 km), both domes having indications of a summit crater, resolvable in a 200 mm aperture. To the west, the low illumination reveals wrinkle ridges, more lumpy dome-like elevations, and the buried crater Wallace (26 km) – much of whose sharp rim has not been covered by Imbrium lava flows. Between Wallace and Mons Wolff there is a very thin, sharp ridge about 80 km long. A sharp, bright little crater perches on the Apennine front, its light-coloured ejecta dusting the mare surface, readily visible at low illumination.

Montes Apenninus is the Moon's grandest mountain range, and there is no better time to appreciate its grandeur than during day eight. Displaying clear radial sculpting in the arrangement of its mountains and valleys, the clear-cut border with Mare Imbrium exhibits several prominent headlands that have been allocated mountain names – Mons Bradley, Mons Huygens, Mons Ampère and Mons Wolff. From peaks on the Apennine front which tower to heights of 5000 metres or more, the range gradually diminishes in height towards Mare Vaporum. The sharp-rimmed crater Conon (22 km) is embedded to a depth of 2300 metres in the Apennines just east of Mons Bradley.

Farther south, the sinuous Rima Conon winds for 45 km down the valley of Sinus Fidei (Bay of Faith) on the northern shore of Mare Vaporum. An exceedingly eroded crater, Marco Polo (25 km), can just be picked out from among the jumbled mountain mass off Mare Vaporum's north-western coast

An extensive though rather bland hilly upland region resembling a blanket of moss separates Mare Vaporum from Sinus Aestuum (Bay of Billows). The south-eastern part of Sinus Aestuum is crossed by a prominent braided wrinkle ridge. To the north can be traced the eastern wall of Eratosthenes (58 km) and a few of its radial eastern ridges – a great crater that is about to be revealed in all its glory (see 'Day nine'). In the uplands north of Sinus Medii, the linked double plain of Pallas (50 km) and Murchison (58 km) is an easily identifiable target, despite its rather eroded nature. Pallas, in the west, has a sizeable central mountain, and its eastern wall merges with Murchison, whose south-eastern wall in turn has been completely flooded with the lavas of Sinus Medii. Just north of Pallas, the crater Bode (19 km) has an intriguing tongue of material extending from its northern rim on to the small, flat crater floor. North of the crater Bode, Rima Bode is a difficult sinuous rille to resolve, even through a 200 mm aperture. On the other side of the pair, the near-perfect crater bowl of Chladni (14 km) lies on the tip of a ridge extending from Murchison's eastern wall, 50 km west of Triesnecker.

Schröter (35 km), a disintegrated, flooded crater with completely breached southern ramparts, lies some way to the west, marking the eastern border of Mare Insularum (Sea of Islands). Under early-morning illumination, Schröter's eastern rim casts a peculiar shadow on to its flat floor, broad and sharp, angled like the edge of a modelling knife. Between Schröter and the western edge of Sinus Medii lies a wide (currently unnamed) plain whose mottled surface is punctuated with several small rounded hills and a small wrinkle ridge. Rima Schröter runs 40 km south to the bay crater Sömmering (28 km), but it is a very difficult object to discern, even in large instruments. Nearby, the crater Mösting (26 km) has a sharp rim with a V-shaped kink in the north, its inner walls displaying prominent bench-like terracing. The 2700-metre-deep bowl crater Mösting A (13 km) lies on the western wall of Flammarion (75 km), south of Sinus Medii. Mösting A's prominence and near-central position on the lunar disk has led to it being designated the fundamental point in determining the Moon's network of coordinates. Flammarion contains half a dozen low domes, and its northern edge is crossed by Rima Flammarion, a narrow rille 80 km long and visible through 150 mm apertures.

Dominating the south central portion of the day-eight terminator are the magnificent walled plains Ptolemaeus (153 km), Alphonsus (118 km), Arzachel (97 km), Purbach (118 km) and Regiomontanus (126 × 110 km), all easily visible in binoculars. Of these walled plains,

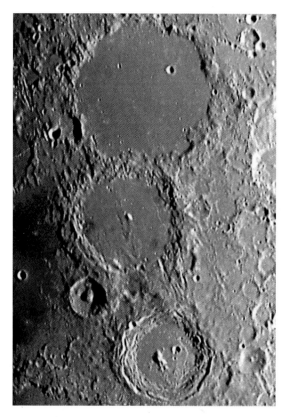

► *Walled plains near the centre of the lunar disk – (from top) Ptolemaeus, Alphonsus and Arzachel. The image was obtained on 24 April 1999 by Mike Brown, using a 370 mm reflector and a HX516 CCD camera.*

Ptolemaeus is the prime attraction for most cursory visual lunar observers. Ptolemaeus demands attention by virtue of its sheer size, even though it is not the most geologically varied feature on the Moon. It has a somewhat polygonal outline, and its floor, bounded by uncomplicated walls, lies 2400 metres below the rim. The floor appears smooth and relatively featureless in a 60 mm telescope (even when illuminated by a low Sun), save for the sharp little bowl crater Ammonius (9 km) in the north-eastern sector of its floor. On the crater's floor, visible through a 150 mm telescope, are Ptolemaeus B, a 15 km diameter ghost ring adjoining Ammonius in the north, along with an abundance of shallow, rimless dimple craters and a smattering of tiny craterlets on the verge of resolution.

The region around Ptolemaeus – quite noticeably in the north and west – has been sculpted into a fantastic linear pattern by the immense forces unleashed by the asteroidal impact that formed the Imbrium basin. The lovely circular Herschel (41 km), a crater with a sharp rim, lies just north of Ptolemaeus. It has prominent internal terracing, and its floor is sunk some 3770 metres beneath the crater rim, watched over by a small central mountain. To the south-west lies Ptolemaeus E, a deformed, elongated flooded crater. Davy (35 km), 100 km to the west of Ptolemaeus, lies on the eastern border of a large unnamed lava plain, a northern extension of Mare Nubium (Sea of Clouds). A large bowl-shaped crater, sharp and polygonal with a flat, mottled floor, sits on its south-eastern rim. Davy itself perches on the southern wall of a much larger, distinctly rectangular walled plain with disintegrated walls. The plain is notable for Catena Davy, a line of little craters crossing eastwards from the centre of the plain to a bright little crater on the eastern rim. If this crater chain is a line of secondary impacts, it is not immediately evident where they have come from. A large flooded bay in the northern ramparts of Davy Y is formed by Palisa (33 km).

Adjoining Ptolemaeus' southern wall, Alphonsus makes a topographically fascinating subject that will reward high-magnification perusal with large apertures. Alphonsus' floor is mottled and bears wrinkles which follow the lines of the Imbrium sculpture, suggesting that the lava fill is not as deep as it is in Ptolemaeus. There are several interesting craterlets with dark haloes, thought to be a strong indication that the craterlets are volcanic vents, since they clearly lie along several narrow fault valleys. Alphonsus' central peak is a compact pyramid that rises to a height of around 1500 metres; it lies on a prominent line of ridges which bisects the crater from north to south. A 200 mm telescope will resolve Rimae Alphonsus, a narrow rille network on the eastern part of the floor.

To the south-west of Alphonsus can be found Alpetragius (40 km), a prominent, sharp-rimmed polygonal crater with a central elevation of gigantic proportions – a great rounded mountain with a north–south base one-third of the crater's diameter and rising to about 2000 metres. Volcanic eruptions may have added a great deal of mass to the original central uplift, and there are indications of an eroded vent on its summit. The ruined, flooded crater Alpetragius X (33 km) immediately to the west provides a nice contrast. A short distance south-east across the uplands, Arzachel makes a superb addition to the 'Ptolemaeus chain'. Arzachel is a beautifully formed impact crater, with highly developed walls and internal terracing, a large central elevation offset to the west, and a well-developed collar of ridges and furrows. Arzachel's floor contains several large craterlets and a cleft which winds along the floor from the north to the south-east, visible in a 200 mm aperture.

About 100 km farther south, the conjoined plains of Purbach and Regiomontanus mark the beginnings of the lunar southern uplands proper. Both plains are substantially eroded, with broken walls, craterlets and scattered peaks spread across their floors. Purbach in the north is the more well-defined of the two, with a sharp but somewhat dented eastern rim and a more chaotic western margin. A teardrop-shaped plain, Purbach G, intrudes upon the northern rim, and south of this, spreading across Purbach's floor, are a collection of hills and peaks, the remains of Purbach W near the centre of the plain, and the sharp little bowl of Purbach A to its south. A 200 mm instrument reveals a mass of fine craters on the flatter, eastern part of Purbach's floor, and to the west, under early-morning illumination, can be seen a prominent gorge, not sharp enough to be called a rille, which winds for 40 km northwards across the plain. A group of mountains and several crater indentations lie along the boundary between Purbach and Regiomontanus to the south. Regiomontanus is irregular in outline and noticeably elongated east–west. Its wall is eroded, highly scalloped and pitted with craters. The eye is drawn to a large (25 km long) mountain massif on its northern floor, topped by a neat little bowl crater – possibly a volcanic vent, although its freshness compared with the rest of Regiomontanus makes it more likely to be an impact feature.

South-west of Regiomontanus, the vast ancient plain of Deslandres (234 km) is slowly coming into view, a wide shadow being cast by its eastern rim, indented by Walter to its east. Deslandres' floor is a bewildering mass of hills and little craters. But there seems to be a

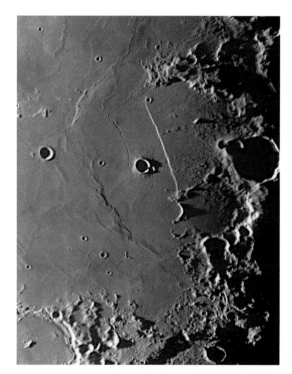

▶ *Rupes Recta, a large fault in Mare Nubium. The image was obtained on 21 September 2000 by Mike Brown, using a 370 mm reflector and a HX516 CCD camera.*

subtle order underlying the apparent confusion, for there can be traced in the rough terrain the vague outlines of several large buried craters of great antiquity. A beautiful linked chain of six craters, visible in a 60 mm telescope, traverses 35 km of the eastern part of the floor from north to south. Deslandres' southern edge is dented by a large, prominent bay formed by the crater Lexell (63 km), whose northern edge has been breached and buried by lava flows. The mid-western sector of the floor is dominated by the crater Hell (33 km), with its sharp rim, and its floor 2200 metres deep which contains a small but prominent mountain. Crossing the western flanks of Hell is another prominent crater chain, aligned in the same direction to the one in the east but twice as long. It is likely that the two chains represent secondary cratering from the debris blasted out by the Tycho impact, a couple of hundred kilometres to the south. Deslandres' western wall is dotted with some sizeable craters, the most intact section of it lying in the north-west.

As the terminator draws back, the eastern plains of Mare Nubium come into view. A large bay in the south-eastern border of Mare Nubium is outlined in the north by the rough mountain mass of Promontorium Taenarium. Under early-morning illumination the bay appears to be part of a large flooded basin about 200 km across, its buried western rim outlined by an intricate complex of wrinkle ridges. The eastern border is home to Thebit (57 km), a prominent crater, with terraced internal walls and a flattish wrinkled floor with a small semicircular trough (visible in a 200 mm aperture). Thebit's southern rim has a couple of indentations, and its western rim is overlain by the sharp crater Thebit A, whose own western rim is in turn overlain by a smaller crater, Thebit L. Close inspection of Thebit L with a 150 mm aperture will reveal a tiny central peak.

The main attraction of the marial plain west of Thebit is Rupes Recta (Straight Wall), a large, slightly curving fault 110 km long. The shadow it casts is visible even through binoculars as a prominent black line at morning illumination. It is the neatest and most clearly visible product of faulting visible on the entire Moon. Thebit D, a small crater, lies at the fault's northern tip. To its south lie the Stag's Horn Mountains (unofficial name), a collection of peaks rising up to 1000 metres from the plain. They are evidently the remains of the western wall of a buried crater which extends to the mare border, indenting it south of Thebit. Immediately to the west of Rupes Recta is the bright, sharp bowl crater Birt (17 km), with Birt A (7 km) lying on its eastern flanks. West of Birt, running 50 km north from the craterlet Birt F to Birt E, a narrow rille can be resolved with a 150 mm aperture. Unlike Rupes Recta, this rille is not the product of faulting – it is a slightly sinuous channel, cut by the erosive forces of extremely ductile lava around 3 billion years ago.

Day nine

Day eight's terminator was memorable for the grandeur of Montes Apenninus and the giant walled plains in the southern central regions of the disk. Day nine's terminator opens up some widely separated grand lunar showpieces. With the unaided eye, the eastern maria are clearly discernible. In the north, the undulating line of Mare Frigoris continues to be uncovered, the terminator bisects Mare Imbrium, and Mare Nubium has fully emerged into the morning sunlight.

Binoculars show several major features near the terminator – at the western tip of the lunar Apennines, Eratosthenes is prominent. Some 250 km to its south-west, Copernicus (93 km), one of the Moon's most imposing craters, is coming into view, complete with its fabulous ray system. In the southern uplands, Tycho (85 km), another major crater and centre of a prominent ray system, can be seen, and 340 km to its south the vast walled plain of Clavius (225 km) is coming into view. The brilliance of Copernicus and Tycho with their ray systems are obvious to the naked eye, but some keen-sighted observers claim to have discerned the dent in the terminator caused by Clavius – and this presents a target of less than 2'.

Even a 60 mm telescope shows the walled plains north of Mare Frigoris to good effect, with the bright Anaxagoras and its rays an ever-prominent feature. On the northern border of Mare Frigoris, the crater Fontenelle (38 km) has a bright rim, and the interior shadows gradually roll back to uncover a flat, hummocky floor. To the east lies a scattered cluster of island peaks beneath which a wrinkle ridge casts a broad shadow; 40 km south of Fontenelle lies a tiny unnamed crater surrounded by a brilliant white ejecta collar – the Linné of the far north. Mare Frigoris, up to the terminator, is undulating terrain strewn with tiny peaks, craterlets and wrinkle ridges. Later in the day, the craters La Condamine (37 km), on the southern shore of Mare Frigoris, and Maupertuis (46 km), to its south in the western Montes Alpes, come into view. La Condamine is slightly irregular in outline – a 200 mm aperture may reveal the tiny bifurcated rille on its knobbly floor. Maupertuis is considerably more eroded, with a battered rectangular outline and a floor crossed by multiple ridges. To its west, Rimae Maupertuis represent a significant challenge for a 200 mm telescope. Farther west begins the spectacular range of Montes Jura, marking the northern boundary of Sinus Iridum (Bay of Rainbows) in the northern Imbrium plain. During the ninth day the eastern parts of Montes Jura, including the south-pointing wedge of Promontorium Laplace, are becoming increasingly eye-catching. In northern Mare Imbrium, between Promontorium Laplace and Montes Teneriffe, lie

Montes Recti (Straight Mountains), one of the Moon's more unusual ranges. Measuring 80 km long, averaging 20 km wide and aligned east–west, the range has peaks that rise to 1800 metres. A small crater is located near the eastern tip of the range. Montes Recti are part of a largely buried inner ring of the Imbrium basin, part of the same underlying structure that links them with Montes Teneriffe and Mons Piton in the far east.

Helicon (25 km) and Le Verrier (20 km), two prominent sharp-rimmed craters, lie in the north-western sector of Mare Imbrium, 100 km south of Promontorium Laplace, 30 km apart. The pair's

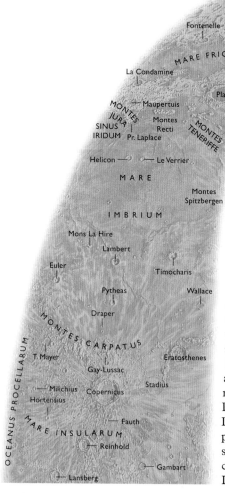

external and internal topography is almost identical. A low Sun brings to prominence several large wrinkle ridges to the east, winding their way south across the Imbrium plain from the vicinity of Montes Teneriffe. Dorsum Grabau, a prominent wrinkle ridge, extends for 120 km from Montes Spitzbergen, south of the tiny crater Landsteiner (6 km) to join with Dorsum Higazy west of the prominent crater Timocharis (34 km). Timocharis is a sharp-rimmed, slightly polygonal crater, with intricate radial ridges and clearly defined internal terracing. Instead of a central peak, Timocharis sports a crater at the centre of its floor. Just over 200 km to its west lies a very similar though slightly smaller crater, Lambert (30 km), whose floor also displays a central crater. Several wrinkle ridges converge on Lambert – Dorsum Zirkel to the north-west, Dorsa Stile to the north-east and a prominent group of ridges to the south, several of which mark the circular outline of a buried crater, Lambert R (45 km). Farther south is the sharp polygonal crater Pytheas (20 km), which is connected to the tiny bowl-shaped Draper (9 km) by a narrow braided wrinkle ridge. A mountain island, Mons La Hire (20 km long and 10 km broad), lies 125 km from Lambert's north-western rim – at high magnification the mountain looks a little like Montes Recti, even down to the crater that has been planted on its eastern heights. When Mons La Hire is on the morning terminator, scan the plains 40 km to its south and try to spot the tiny craters Artemis and Verne, both 2 km in diameter – a challenge for a 150 mm aperture.

Eratosthenes is a superb object near the western extremity of Montes Apenninus, a well-developed impact crater with complex interior terracing and a group of central mountains. Although there is a complete absence of a ray system, clear secondary impact structure can be traced in the surrounding terrain, with prominent concentric ridges that give way to narrow secondary impact grooves, some of which extend more than 100 km from Eratosthenes' rim. To the south lies Stadius (69 km), an unusual circular feature with low, slender walls, broken in places by tiny craters. Stadius' north-eastern wall is linked to Eratosthenes by a rounded group of hills. To the west, the morning terminator early in day nine uncovers some wonderful sculpted valleys and numerous winding chain craters – topography that heralds sunrise over their source, the mighty crater Copernicus.

Copernicus is one the finest impact features in the entire Solar System. The bright rays which surround it in all directions can easily be seen with the naked eye just beneath Mare Imbrium when illuminated by a mid-morning to early-evening Sun. Magnified in a

◄ *Mighty Copernicus, under a morning Sun. The image was obtained on 4 March 2001 by Mike Brown, using a 370 mm reflector and a HX516 CCD camera.*

telescope eyepiece, Copernicus and its environs are thrilling to observe. Huge amounts of debris were flung out from the impact, some 900 million years ago, the heftier fragments producing secondary impact craters, grooves and crater chains arranged radially around Copernicus. In places, the piles of ejected material are many tens of metres thick.

Copernicus' floor lies 3760 metres beneath its rim. The southern parts of the floor are rougher than the northern, but small telescopes cannot resolve this terrain, which in a 60 mm aperture appear smooth but darker in tone than in the north. A group of mountains occupies the centre of the floor, the highest of them rising to 1200 metres. Complicated terracing adorns Copernicus' inner walls – indeed, so much detail on the inner west wall can be seen through a 100 mm telescope that only a brave and highly proficient observer would ever attempt to draw the whole lot in a single observing session. Copernicus A (2.5 km) sits snugly on the middle terrace due east of the central peaks, and to its east the rim displays a marked kink protruding towards the craterlet.

Copernicus' rim rises 900 metres above the surrounding terrain. An exceedingly intricate network of concentric hills, furrows and radial ridges comprises the outer slopes. The north-western and western slopes display a distinct boundary between the concentric features and the radial ridges, around 20 km from Copernicus' rim, forming a ledge that casts a prominent westward shadow during the early morning. This concentric topography was formed when the shock waves of the original impact were frozen into the rock and later modified by landslips, and it gives Copernicus the impression of having a substantial double wall. The radial ridges farther from the rim are composed of massive piles of ejected material. About 100 km to the north, Montes Carpatus (Carpathian Mountains) mark part of the southern boundary of Mare Imbrium; they form a broad but jagged east–west arc about 400 km long, with peaks that rise on average between 1000 and 2000 metres. Embedded in the southern Carpathians is the disintegrated crater Gay-Lussac (26 km) and a prominent rille to its west. On the other side of Copernicus, 50 km from its southern rim, lies the keyhole-shaped double crater Fauth (12 km) and Fauth A (10 km), its orientation – pointing directly away from Copernicus – suggesting that it is the product of a major secondary cratering from the impact that formed Copernicus.

Mare Insularum is an appropriate name for a patchy lava plain containing a multitude of isolated peaks, hills and dome-like elevations spread over tens of thousands of square kilometres to the south of Copernicus. The north-west sector of Mare Insularum contains the most extensive dome fields on the Moon, dozens of them accessible with a 100 mm telescope. Immediately south of the flat-floored crater Tobias Mayer (33 km) is a cluster of around a dozen undesignated domes, several of which are topped by tiny summit craterlets. A very large semicircular swelling, about 40 km in diameter, can be seen to their south. The scattering of small peaks and hills in among these domes makes the area particularly attractive at high magnification, and some of the domes play hide-and-seek with the shadows of these higher features. A neat circular dome with a 10 km base and a summit craterlet lies to the west of the well-defined bowl of Milichius (13 km). To the east, the morning Sun brings to light a prominent north-west/south-east line of six mountain peaks, more than 100 km long. These may be the remains of the western rim of an ancient buried crater, further traces of which are visible at a low angle of illumination west of Copernicus, including what may be the scattered remnants of a central elevation. Farther south, the sharp bowl crater Hortensius (15 km) has its own clearly visible cluster of half a dozen sizeable domes, most of which have summit craterlets.

Gambart (25 km), with its simple, clear-cut wall and smooth flooded floor, is a notable feature in southern Mare Insularum. Reinhold (48 km), 220 km to its west, is a prominent crater with external ridges, internal terracing and a rather flat floor with one or two tiny hills. To its north-east, the hexagonal plain Reinhold B (25 km) has a flat floor depressed beneath the mean level of the mare. Lansberg (39 km), on the south-western margin of Mare Insularum, is a classic impact structure, with clearly defined terracing and a multiple central mountain. Montes Riphaeus (the Riphaeus Mountains – an ancient name for the

Ural Mountains) is a prominent range, 150 km long, whose several branches give it the appearance of a waterfowl's footprint, complete with webs. Another, less prominent and unnamed mountain range lies to the west of Montes Riphaeus, and between the two lies the bright sharp bowl of Euclides (12 km) with its small ray system.

Fra Mauro (95 km), Bonpland (60 km) and Parry (48 km) are a tightly welded trio of walled plains south-east of Mare Insularum. A section of Fra Mauro's eastern wall is missing, having been overrun by lava flows from the plains to the east. Its northern ramparts are low and heavily striated, and this north–south sculpting structure can be traced across the crater's western floor. A rille traverses the terrain north of Fra Mauro, crosses the crater's northern wall and bisects the crater floor. It then splits into two on the crater's southern plain – one branch cuts across the crater wall into Bonpland, going a little beyond its southern ramparts, and the other crosses the wall on to the far western part of Parry's floor and beyond, southwards to the west of disintegrated Guericke (58 km), and is overlain by Tolansky (13 km). A couple of other rilles cut across Bonpland and Parry to the south.

Late on day nine the terminator passes over Bullialdus (61 km), an imposing feature in western Mare Nubium. When this crater is emerging from the terminator, prominent radial ridges can be seen to stretch across the mare surface for up to 50 km. The rims of Bullialdus and two other large craters to the immediate south (Bullialdus A and B) are brightly illuminated. East of Bullialdus lies an interesting line of six large craters in various states of burial, including (from north to south) Opelt (49 km), Gould (34 km) and Wolf (25 km). As the Sun rises, Bullialdus takes centre stage in this region of the Moon. Its rim is sharp and a little irregular, and on its wide inner walls is some of the finest terracing to be seen in any crater. Bullialdus' floor, 3510 metres beneath its rim, is rather dark and slightly convex with a prominent central mountain massif.

Interesting features in the region of Bullialdus include: Lubiniezky (44 km), to the north, a flooded crater with a narrow rim; König (23 km), to the south-west, a sharp rimmed crater with a lumpy floor; and Kies (44 km), a flooded crater with a low, broken rim in the north and a finger-like projection of south wall. West of Kies lies Kies Pi, one of the Moon's loveliest domes, with a base 17 km across and topped by a summit craterlet that is resolvable in a 150 mm telescope.

Campanus (48 km) and Mercator (47 km) are like a pair of giant handcuffs linking south-western Mare Nubium to Palus Epidemiarum (Marsh of Epidemics). A rille cuts between the two, joining the two marial areas and drawing the eye south towards the sharp-edged, flat-floored crater Ramsden (25 km) and an intricate network of rilles nearby – Rimae Ramsden. Visible in a 100 mm telescope, the half dozen or so interlocking rilles measure from about 20 to 50 km in length. At the northern tip of Rimae Ramsden lies Marth (7 km), an intriguing crater within a crater, resolvable in a 200 mm aperture. Rima Hesiodus to the east slices through the lunar crust for 300 km from Palus Epidemiarum across Rupes Mercator and into southern Mare Nubium – an easy object for a 100 mm telescope. Capuanus (60 km), on the southern border of Palus Epidemiarum, is a walled plain with an irregular, battered wall. A 100 mm aperture will resolve several dome-like swellings on its mottled dark grey floor.

Pitatus (97 km) is a large, impressive walled plain on Mare Nubium's southern border. Hexagonal in outline, it has a complicated wall, embossed with craters, with a complex hilly southern inner terrace. A prominent offset peak rises above its mottled floor, stained here and there with darker lava patches, but what makes Pitatus so intriguing is its internal system of rilles, which meander around the periphery of the floor near the base of its internal walls.

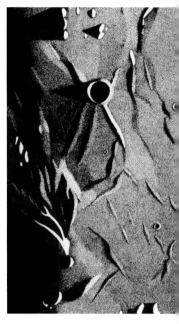

▲ *Wrinkle ridges in Mare Nubium near Nicollet. This observational drawing was made on 22 November 1993 by Harold Hill, using a 209 mm SCT.*

Shining like a beacon in the southern uplands, Tycho (85 km) is the Moon's greatest ray crater. Formed by an asteroidal impact around 100 million years ago, its floor is nearly 5 km below the level of the clean-cut rim, and its bright central peaks rise to 1600 metres above its floor. Tycho looks remarkably fresh. Components of its impressive ray system radiate for more than 1000 km, and though they look thin and translucent, in places the rays are massive piles of debris many tens of metres deep.

Numerous very large, imposing walled plains are on view in the southern uplands, including Wilhelm (107 km), the prominent Longomontanus (145 km) with its eastern rim overlapping a smaller crater, Longomontanus Z, and the massive Maginus (163 km) with its wide blocky walls. But the showpiece of the southern uplands, visible in all its glory on day nine, is Clavius, a feature of grand proportions and great beauty. Fully 225 km in diameter, Clavius has a sharp, scalloped rim in which several large craters reside – Porter (52 km) on the northern wall, Clavius L and K in the south-west, and Rutherfurd (50 km), which straddles Clavius' inner wall up to the south-eastern rim. Both Porter and Rutherfurd have associated radial impact ridges which, though subsequently overlain by lava flooding in Clavius' interior, remain visible. Both have sizeable mountains on their floors, but those of Rutherfurd are the more prominent, with one particularly large mountain block to the north. Through a 60 mm telescope the floor of Clavius is a smooth grey expanse interrupted only by an arc of craters of diminishing diameter that springs from Rutherfurd and proceeds to the west, through Clavius D (27 km), Clavius C (22 km), Clavius N (11 km) and Clavius J (10 km). A 200 mm aperture will reveal up to 30 craters on Clavius' floor, and its internal western wall resolves into a spectacular, roughly chiselled semicircle shining brightly in the morning Sun. North of Clavius can be seen some of the original radial impact sculpture, though it has been considerably eroded over time.

Scheiner (110 km) and Blancanus (105 km), two large walled plains beyond the south-western ramparts of Clavius, are prominent features near the morning terminator, their deep, dark floors both pitted with small craters. To their south, the linked crater pair Klaproth (119 km) and Casatus (111 km) are less prominent, and as the shadows recede to reveal most of Klaproth's smooth dark floor, the interior of Casatus remains deeply shadowed, but eventually the sharp circular crater Casatus C is revealed to the north of its flat floor. Closer to the southern lunar limb, Moretus (114 km) is a nicely defined impact crater embedded among the crater fields of the southern uplands. Moretus' inner western wall shines brightly in the morning sunshine, displaying abundant terracing structure and a prominent single peak, which casts a pointed shadow on to its smooth grey floor. Short (71 km), Newton (79 km) and Cabeus (98 km) are three near-limb craters that anyone wishing to familiarize themselves with the topography of the south polar region should get to know. Their degree of foreshortening is a useful indicator of the extent of libration and whether features near the lunar south pole are on view.

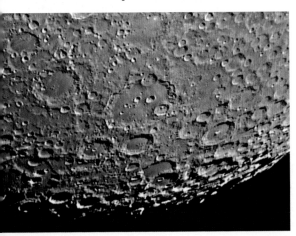

◀ The crater-crowded lunar southern uplands, with Clavius at centre. The image was obtained on 23 April 2002 by Brian Jeffrey, using a 102 mm achromat with a Philips ToUcam Pro CCD camera.

Day ten

DAY 10

With the unaided eye, almost the whole of Mare Imbrium can now be discerned, including a small irregularity in its northern margin formed by the bay of Sinus Iridum and sharply outlined in the north by the steep, bright arc of Montes Jura. The bright rays around Copernicus and Tycho are now easy to spot without optical aid, but it takes excellent eyesight to be able to make out the dark patch of Palus Putredinis in eastern Mare Imbrium and Montes Riphaeus on the south-eastern border of Oceanus Procellarum (Ocean of Storms). Mare Nubium can be seen in its entirety, and emerging from the morning terminator to its west are the eastern plains of the smaller Mare Humorum (Sea of Moisture), which makes a noticeable dent in the southern section of the terminator. Binoculars show a dramatically changing terminator between day ten and day twelve. The remaining heavily cratered highlands on the terminator lie in the far north and south, but much of the terminator in between sweeps across relatively flat marial plains.

Philolaus (71 km), a prominent crater in the northern highlands, covers most of a larger, more eroded crater, Philolaus C, and at sunrise the observer has the impression of an extensive single crater with a

▶ *The ten-day-old Moon in an image obtained on 11 May 2003 by Peter Grego, using a 150 mm achromat and a Ricoh RDC-5000 digicam.*

large curving central mountain (actually the western rim of Philolaus). Visible at a higher angle of illumination are prominent interior terracing and two mountain blocks on Philolaus' floor. South-east of Philolaus is a large, well-formed walled plain some 100 km in diameter; currently unnamed, it is one of the largest clearly discernible craters on the nearside without a proper designation. It joins a smaller 70 km diameter plain to the south, giving it a squashed keyhole shape. Farther along the terminator, the irregular, disintegrated eastern wall of J. Herschel (156 km) is on view, its floor largely in shadow and just a few areas of high relief catching the sunlight.

Mare Frigoris' western reaches are now becoming visible, the low Sun bringing out lots of small isolated mountains, hills and craters. Guarding the mare's entrance at the junction with Sinus Roris (Bay of Dew) is Harpalus (39 km), a prominent polygonal crater with internal terracing and a cluster of small hills on its floor. Harpalus is the centre of a small ray system. To the south, the sweeping arc of Montes Jura is at its most impressive, from the broad, blunt east-pointing cape of Promontorium Heraclides in the west to the south-pointing wedge of Promontorium Laplace. The mountains rise steeply from the plains of Sinus Iridum, reaching heights of 6000 metres. The prominent, clear-cut craters Sharp (40 km) and Bianchini (38 km) have been gouged out of Montes Jura; each is slightly polygonal in outline with internal terracing, a flattish floor and a central elevation. At this morning illumination it is clear that the range marks the northern wall of a large buried

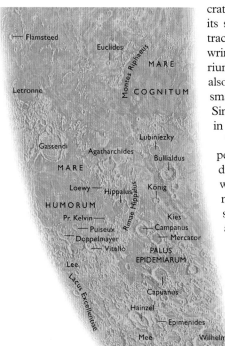

crater about 250 km across, its submerged southern rim traceable in a number of wrinkle ridges on the Imbrium plain to the south. It is also possible to make out a smaller inner ring within Sinus Iridum, some 120 km in diameter.

The unnamed upland peninsula west of Sinus Iridum is rough and dotted with a myriad of undesignated craters. The large sharp-rimmed crater Mairan (40 km) dominates the area. Two huge, rounded mountain blocks, Mons Gruithuisen Gamma and Mons Gruithuisen Delta, rise from the north-western Imbrium shoreline. Each rises to nearly 1000 metres and has a base 20 km across. Gamma is the more regular of the pair, dome-like and topped with a summit crater visible in a 100 mm telescope, though its volcanic origin is debatable. To the south lies the crater Gruithuisen (16 km), around which are scattered several isolated highland clumps and the northern part of the Dorsum Blucher wrinkle ridge complex. Delisle (25 km) and Diophantus (19 km), along with the shark's tooth of Mons Delisle, make a lovely group for close scrutiny, though the narrow sinuous Rima Diophantus (140 km) may be beyond the capabilities of a 200 mm aperture. Farther south can be seen the prominent mountain group of Mons Vinogradov, Euler (28 km) to the east, and rough terrain to the south containing the inconspicuous little craters Jehan (5 km) and Natasha (12 km).

Sitting at the centre of its prominent starburst of rays, the crater Kepler (32 km) is now emerging from the terminator, some 480 km west of Copernicus. The rays can be traced for several hundred kilome-

▲ *Northern Mare Imbrium, from Plato to Sinus Iridum and the Montes Jura. The image was obtained on 23 April 2002 by Brian Jeffrey, using a 102 mm achromat and a Philips ToUcam Pro CCD camera.*

tres in all directions, mingling with Copernicus' rays in Mare Insularum. For all the splendour of its rays, Kepler is an understated crater with a low, sharp rim, a rather restrained system of concentric ridges, very little terracing on its thin inner walls, and a bumpy floor devoid of a central peak. A spray of elevations emerges to the north of Kepler, including a 15 km diameter dome, which is an easy object through a 100 mm aperture. Encke (29 km), a distinctly polygonal crater, lies 85 km south of Kepler, and like its neighbour has a hilly floor. It lies on the disintegrated north-eastern wall of a large plain, Encke T (100 km), clearly visible during the early morning. To its west lies the little crater Maestlin (7 km), a couple of buried craters, and Rimae Maestlin (80 km long). Rimae Maestlin are a challenge for the user of a 150 mm telescope.

Many long wrinkle ridges can be seen flowing north–south down the mid-eastern part of Oceanus Procellarum from the general vicinity of Encke to the southern mare border, and the remnants of numerous submerged craters are traceable in the light of the early-morning Sun, including Flamsteed P (95 km), where the first US lunar soft-lander, Surveyor 1, touched down in 1966. South of Flamsteed (21 km), Dorsa Rubey extends into a large bay crater on the southern edge of Procellarum. Letronne (119 km) is a crater whose northern ramparts have been obliterated by lava flows, but the small group of hills at its centre are undoubtedly the remnants of its central massif.

Gassendi (110 km) is the glorious gem of the south. This grand walled plain on the northern border of Mare Humorum is one of the Moon's most beautiful craters. Circular in outline, the walls of Gassendi rise to a height of 3600 metres in places, but they are considerably lower in the south, where they dip down to almost the level of Mare Humorum. Gassendi's northern rim is cut across by the sharp polygonal crater Gassendi A (33 km), and this in turn touches Gassendi B (25 km). Gassendi's floor is very complex. Half a dozen central mountain peaks, two of which are particularly large, rise to nearly 1000 metres above the floor. In a 60 mm telescope, the floor surrounding the peaks appears mottled and lumpy, but a 150 mm aperture resolves it into many narrow rilles and low peripheral ridges.

In the early morning, a broad black triangle of shadow is cast over the northern part of the floor between Gassendi A and a low ridge to its south, and a bright scarp face veers off from the western wall, jutting into the black shadow beyond.

Mare Humorum is a beautiful, near-circular sea with an east–west diameter of 380 km. It has a well-preserved mountain border, except in the north-east where it links with the plains of southern Oceanus Procellarum. By the end of day ten Mare Humorum has fully emerged from the morning terminator. Binoculars show a dark, smooth surface, but a high-magnification telescopic view reveals numerous sharp craters dotted here and there, along with a network of concentric wrinkle ridges that appear particularly prominent in the east. Partially submerged craters dot the uplands on its eastern border, notably the flooded plains of Agatharchides (49 km) and Loewy (25 km). A bay in the inner shoreline of the mare is formed by Hippalus (58 km), through which cuts one of the deep, prominent rilles of Rimae Hippalus, visible in a 60 mm telescope. Concentric to the border of Mare Humorum, the five main rilles in the system stretch up to 250 km and measure 3 km across in places. Nearby, rising above the eastern plains of Mare Humorum, Promontorium Kelvin is an impressive-looking mountain with a base measuring 40 × 30 km, rising to 2000 metres above the mare. Vitello (42 km), a low-walled crater with a sharp rim and a hilly, clefted floor, lies on the mare's southern shore. To its west can be seen the ghost crater Puiseux (25 km) and the large, partially flooded crater Doppelmayer (64 km), whose northern wall has been overwhelmed with mare lava flows, but a prominent central peak and number of low ridges on its floor remain clearly visible. Mare Humorum's western shoreline is sliced through by a number of rilles, including Rimae Mersenius, Rimae Palmieri and Rimae Doppelmayer, along with some prominent undesignated rilles in the highlands to the north-west.

A small, dark, irregularly shaped lava plain, Lacus Excellentiae (Lake of Excellence), lies to the south of Mare Humorum. One of the more notable features near the southern uplands terminator during day ten is the large, considerably eroded crater Mee (132 km), with a trio of deformed, conjoined craters on its northern wall – Hainzel (70 km), Hainzel A and Hainzel C. To its east is the small, elongated plain of Lacus Timoris (Lake of Fear, 130 × 50 km). Schiller, a prominent, somewhat elongated walled plain (171 × 71 km), comes into view in the south-east. Possibly a line of three joined craters (perhaps resulting from a simultaneous multiple impact), the broader, smoother southern plain of Schiller narrows in the north, where a couple of large ridges cross the floor from north to south. Schiller itself lies on the north-eastern margins of a large but unnamed (350 km diameter) multi-ringed basin, unofficially designated the Schiller Annular Plain. This feature is visible towards the end of day ten and during day eleven.

Day eleven

All the major nearside maria are now visible without optical aid. Binoculars show the smoothest-looking terminator yet, as much of its central portion is taken up by Oceanus Procellarum, but a notable bright spot surrounded by an emerging ray system is visible in its midst. This is the Aristarchus plateau, an area of exceptional geological interest. By the end of day eleven, Mare Frigoris in the north has been fully exposed, and a large dent is made in the southern terminator by the eastern rim of the giant crater Schickard (227 km).

DAY 11

The crater J. Herschel (156 km) has come into view in the north-west, a lovely if somewhat disintegrated walled plain with loose, rubbly walls and an undulating floor. Wrinkle ridges abound along the

J. Herschel

Harpalus

Sharp

Mairan

Gruithuisen

Montes Agricola

Vallis Schröteri

Montes Harbinger

Prinz

Aristarchus

Herodotus

OCEANUS

Rima Marius

Marius

Reiner

PROCELLARUM

terminator in Oceanus Procellarum, a number of which converge on the Aristarchus plateau, a raised, square-shaped hilly area of about 30,000 sq km. A 60 mm telescope shows a slight reddish coloration in this area. One of the brightest features on the Moon, Aristarchus (40 km) is a polygonal crater with steep terraced inner walls, which display dusky radial banding, and a small central mountain at the centre of its floor. Aristarchus' inner walls are among the brightest features of the nearside. Components of the ray system, stretching mainly eastwards of the crater, can be traced for several hundred kilometres.

Immediately west of Aristarchus, Herodotus (35 km) is a dark, flat-floored crater which makes an intriguing contrast with Aristarchus. To the north, Vallis Schröteri – 160 km long, 10 km wide and 1000 metres deep in places – is the finest example of a sinuous rille, and it can be resolved in a 60 mm telescope. Observers call the portion of the valley nearest Herodotus the Cobra's Head because of its distinctive shape. A small dome lies

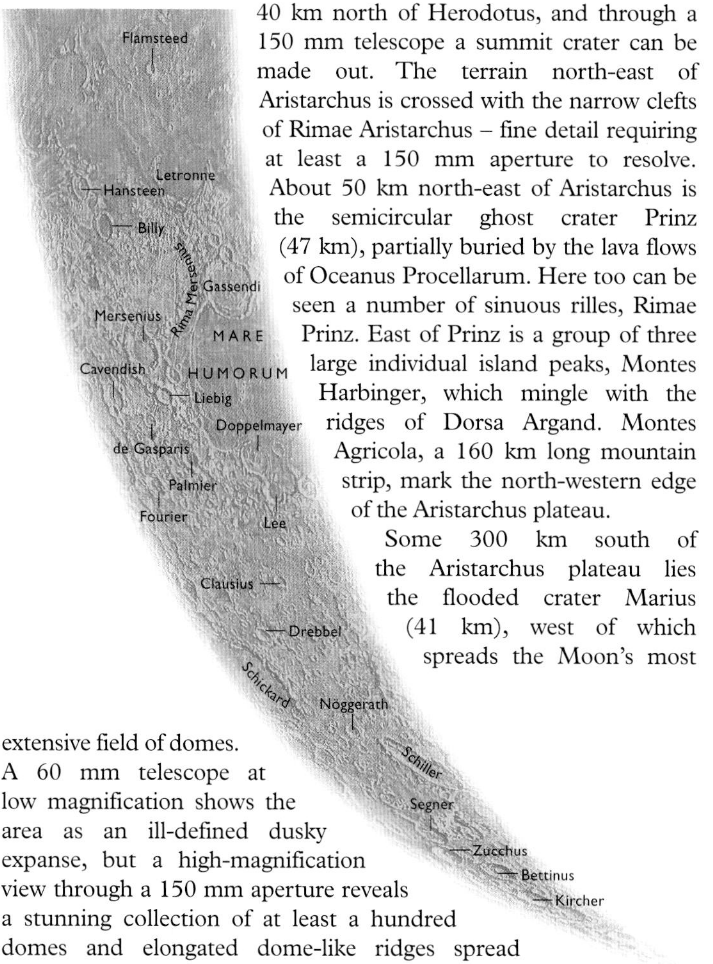

40 km north of Herodotus, and through a 150 mm telescope a summit crater can be made out. The terrain north-east of Aristarchus is crossed with the narrow clefts of Rimae Aristarchus – fine detail requiring at least a 150 mm aperture to resolve. About 50 km north-east of Aristarchus is the semicircular ghost crater Prinz (47 km), partially buried by the lava flows of Oceanus Procellarum. Here too can be seen a number of sinuous rilles, Rimae Prinz. East of Prinz is a group of three large individual island peaks, Montes Harbinger, which mingle with the ridges of Dorsa Argand. Montes Agricola, a 160 km long mountain strip, mark the north-western edge of the Aristarchus plateau.

Some 300 km south of the Aristarchus plateau lies the flooded crater Marius (41 km), west of which spreads the Moon's most extensive field of domes.

A 60 mm telescope at low magnification shows the area as an ill-defined dusky expanse, but a high-magnification view through a 150 mm aperture reveals a stunning collection of at least a hundred domes and elongated dome-like ridges spread across an area of about 40,000 sq km. Wrinkle ridges lead the eye south towards the southern border of Oceanus Procellarum. West of Letronne lies a prominent contrasting crater duo, Hansteen (45 km) and Billy (46 km). Hansteen is polygonal in outline, with traces of external radial furrows and a small rille on its western flanks, internal terracing and a hummocky floor, while Billy is a bland, dark-floored plain. Between the pair, Mons Hansteen is a bright triangular massif some 30 km across.

Mersenius (84 km) is a prominent crater west of Mare Humorum, noteworthy for its distinctly convex floor. The convexity is exaggerated by the crater's generally darker western floor, and a broad, gradually sloping ditch at the base of the inner western wall casting a dark

Flamsteed

Letronne
Hansteen
Billy
Rima Mersenius
Gassendi
Mersenius
Cavendish
MARE HUMORUM
Liebig
Doppelmayer
de Gasparis
Palmier
Fourier
Lee
Clausius
Drebbel
Schickard
Nöggerath
Schiller
Segner
Zucchus
Bettinus
Kircher

shadow at the low angle of illumination. To the south of Mersenius, Cavendish (56 km) is a sharp-rimmed crater with a couple of buried crater rings on its southern floor. It is linked to the inconspicuous crater de Gasparis (30 km) by a system of rilles that cross de Gasparis' floor and extend into the country beyond, including a neat little X-shape in de Gasparis A.

Schickard's eastern wall is making a prominent indentation in the morning terminator late in day eleven, but we have to wait until day twelve to see the walled plain's fascinating floor. Among other near-terminator features of note in the southern uplands is the line of large craters which includes (from north to south) Segner (67 km), Zucchius (64 km), Bettinus (71 km) and Kircher (73 km). Segner and Kircher both have relatively flat, featureless floors, while Zucchius and Bettinus have well-developed internal terracing and central mountains.

Day twelve

All the features visible along the terminator from around day twelve to day sixteen are extremely fore-shortened, and observers must use their imagination to picture how they would appear from a better vantage point. Binoculars will show the huge walled plain of Schickard in the south-west – what a majestic sight it would be if it were closer to the centre of the lunar disk.

DAY 12

Pythagoras (130 km) in the north is a major impact crater with a clear-cut rim and well-developed internal terracing, though we can only observe the inner western wall from the Earth. Its two large central mountains appear almost in profile, the eastern one hiding much of the other from sight – we are, after all, viewing this crater at about 10° to the horizontal. Adjacent to Pythagoras' southern rim lies Babbage (144 km), a walled plain with a somewhat irregular outline, low walls and a smooth grey floor containing a sharp-rimmed 25 km crater. Another unimposing, irregular walled plain, named South (108 km), adjoins Babbage.

In southern Sinus Roris, the Moon's largest dome has emerged from the morning terminator. With a footprint of some 2500 sq km and in places rising to a height of 700 metres, Mons Rümker is an impressive feature, and though it casts a prominent dark shadow to the west at sunrise, its rounded, smooth-sloped topography is evident in that its sunward face is not shining brilliantly, as it would if it were a regular mountain massif. Rümker is by no means a perfect single dome – it is more a lumpy plateau, the remnants of a large complex of lunar volcanoes that was active around 3 billion years ago. The lumpy nature of the feature becomes evident at high powers in a 150 mm aperture – half a dozen blister-like mounds are visible on the plateau. Stretching a couple of hundred kilometres north and south of the Rümker plateau, across the plains of Oceanus Procellarum, are several wrinkle ridges, but the region is one of the blandest on the entire lunar surface. West of the Aristarchus plateau lie Schiaparelli (24 km) and Seleucus (43 km), the latter looking like a scaled-up version of the former – a rather uninteresting, sharp-rimmed crater with a trace of terracing, a flat floor and a small central elevation.

To the south-west of the Marius dome field lie the crater Reiner (30 km) and an unusual feature designated Reiner Gamma, the best example of a lunar 'swirl'. Swirls appear to be delicate, light-coloured surface markings devoid of obvious relief, even at low angles of illumination. Reiner Gamma is a tadpole-shaped smudge measuring about 35 × 100 km, and it is bright enough to be seen easily through binoculars. A 100 mm telescope will resolve some intricate streamers around the feature. Reiner Gamma may have been formed less than 100 million years ago when a small comet impacted on the mare surface, exploding at or near ground level and scouring away the upper layers of dark lunar

▲ *Crater Reiner and the nearby swirl Reiner Gamma form a conspicuous duo in Oceanus Procellarum. From an observation by Peter Grego.*

◀ *The twelve-day-old Moon in an image obtained on 27 March 2002 by Peter Grego, using a 127 mm Maksutov and a Ricoh RDC-5000 digicam.*

soil, exposing lighter soil beneath. West of Reiner Gamma can be traced a small wrinkle ridge that connects with the diminutive crater Galilaei (16 km) to the north.

Complex structure can be observed in the uplands bounding the south-western shore of Oceanus Procellarum. Damoiseau (37 km) is in the centre of a set of concentric ridges and craters, and its own floor is scored with minute clefts, just resolvable in a 200 mm aperture. A large dome-like lobe extends to its north, and to its south are several low connected craters and a set of rilles, Rimae Grimaldi, named after the large crater yet to emerge into the morning Sun. Farther south, Sirsalis (42 km) and Sirsalis A (45 km) are a prominent joined pair of craters in the highlands south-west of Oceanus Procellarum, along with a fine linear rille, Rima Sirsalis, which cuts a 400 km long path from the ocean shore, south of the crater Sirsalis and deep into an undesignated plain east of Darwin, where it intersects Rimae Darwin at right angles.

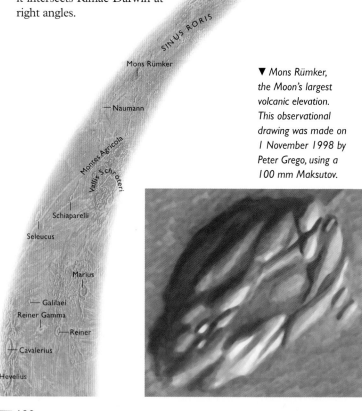

Pythagoras
Babbage
South

SINUS RORIS

Mons Rümker

Naumann

Montes Agricola
Vallis Schröteri

Schiaparelli

Seleucus

Marius

Galilaei
Reiner Gamma
Reiner
Cavalerius

Hevelius

▼ *Mons Rümker, the Moon's largest volcanic elevation. This observational drawing was made on 1 November 1998 by Peter Grego, using a 100 mm Maksutov.*

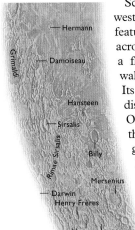

Schickard, a vast walled plain in the south-west, is the most prominent near-terminator feature on view on day twelve. At 227 km across, it is one of the Moon's biggest craters – a flat, circular plain bounded by rounded walls that rise 2500 metres above its floor. Its south-western floor has been noticeably disrupted by debris thrown out by the Orientale impact (more than 1000 km to the west), which has carved elongated grooves, wide ditches and chain craters. Episodes of lava flooding on Schickard's floor have given it its distinctive multi-toned appearance, the floor being darker in the northern and south-eastern sectors. No other flooded lunar crater displays mottling as striking as this. Numerous small craters are dotted about Schickard's floor, at least five of which can be resolved in a 100 mm telescope.

Wargentin (84 km), south-west of Schickard, is an unusual crater which appears to have been filled almost to the brim with lava flows. Several low wrinkle ridges which run across Wargentin's surface are visible in a 100 mm telescope. The craters Nasmyth (77 km) and Phocylides (114 km) stretch south of Wargentin – the rims of the trio are actually inter-locked. Nasmyth intrudes upon the south-eastern wall of Wargentin, and the southern wall of Nasmyth is intruded upon by Phocylides. The comparative lack of craters on Wargentin's surface suggests that it is younger than the surfaces of Phocylides and Nasmyth.

If libration is favourable, the exceptionally large but extremely foreshortened walled plain Bailly (303 km) can be discerned along the southern terminator. An ancient, highly eroded feature, Bailly is actually a multiringed basin, and vague traces of the inner ring (120 km in diameter) are sometimes visible. A number of large craters dot the area in and around Bailly, most notably Bailly A and B on its far southern floor.

Day thirteen

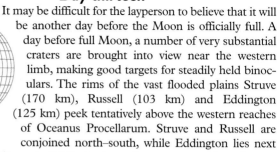

DAY 13

It may be difficult for the layperson to believe that it will be another day before the Moon is officially full. A day before full Moon, a number of very substantial craters are brought into view near the western limb, making good targets for steadily held binoculars. The rims of the vast flooded plains Struve (170 km), Russell (103 km) and Eddington (125 km) peek tentatively above the western reaches of Oceanus Procellarum. Struve and Russell are conjoined north–south, while Eddington lies next to Struve, in the east. Eddington's southern ramparts have been completely obliterated. To its south lie the craters Krafft (51 km) and Cardanus (50 km), whose rims are linked by Catena Krafft, a narrow linear chain crater some 60 km long. A 200 mm aperture may be sufficient to resolve the narrow Rima Cardanus,

running south-west/north-east for 120 km south of Cardanus.

Hevelius (118 km) is a prominent walled plain on the western shore of Oceanus Procellarum, with a broad inner northern wall which displays prominent multiple ridges. Hevelius' generally flat floor is criss-crossed by a system of linear rilles, of which the most prominent two make a neat St Andrew's cross south of the mountain Hevelius Alpha, and are visible with a 150 mm telescope. Hevelius' northern wall links with the deep, prominent crater Cavalerius (64 km), which displays intricate western wall terracing and a central knobbly ridge. Towards the end of day thirteen the large walled plain Hedin, due west of Hevelius, comes into view. Hedin is greatly eroded, with a floor intensely deformed by the effects of the Orientale impact, and like Hevelius it has an associated network of linear rilles.

Farther south can be observed the highly contrasting pair of large walled plains Riccioli (146 km) and Grimaldi (222 km). Seen through a 60 mm telescope,

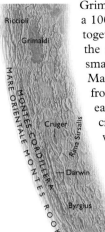

Grimaldi has a smooth, perfectly flat dark floor, but a 100 mm instrument will reveal several low ridges, together with a large but inconspicuous dome on the northern part of the floor, and a well-defined smaller dome near the north-eastern wall. Like Mare Crisium near the eastern limb, when viewed from above, Grimaldi is actually elongated east–west, but this is not obvious because the crater is so foreshortened. Riccioli, to the north-west of Grimaldi, is a sharply defined crater with a complex floor. A smooth, dark lava patch occupies the northern sector of the floor, and half a dozen rilles strike across the brighter, rougher parts of the floor, several going some distance beyond the crater's rim to the north and south.

As the terminator edges ever closer to the edge of the lunar disk, an intricate array of ridges and dark elongated lava patches is revealed along several hundred kilometres of the south-western limb. These are the massive mountain ranges of Montes Cordillera and Montes Rook, which mark the edges of the outer and inner rings of the vast Orientale Basin. Between them lie the lava stains of Lacus Autumni (Lake of Autumn) and Lacus Veris (Lake of Spring). The dark inner patch of Mare Orientale (Eastern Sea) is not yet visible (see 'Day fourteen'). Between Montes Cordillera and Rima Sirsalis lie the small dark lava patches of Lacus Aestatis (Lake of Summer) and the little oval walled plain Crüger (46 km), one of the darkest spots on the Moon. To the south lies Darwin (130 km), an eroded walled plain whose north-eastern floor has a noteworthy dome-like swelling which is cut into by Rimae Darwin. Lagrange (160 km) and Piazzi (101 km), two connected craters near the south-western limb, have battered walls and rough, ridged floors. Nearer the limb are three wide, elongated valleys – Vallis Inghirami (140 km), Vallis Bouvard (280 km) and Vallis Baade (160 km) – formed by secondary impacts from the Orientale event.

Day fourteen – full Moon

Full Moon takes place 14 days and 19 hours after new Moon. During the first half of day fourteen, the last vestiges of shadow in the west gradually diminish in size. In binoculars the lunar disk appears to be 100% illuminated, but a telescope reveals that some full Moons possess a discernible terminator on the northern or southern limb. This happens when the full Moon lies far enough away from the ecliptic plane for the terrestrial observer to see 'above' or 'below' the north or south pole, allowing a glimpse of features on the northern or southern terminator.

Full Moon is time to survey the intricate patchwork of light and dark areas and to marvel at the many ray systems, and binoculars are perfect for both activities. In the absence of shadow to throw them into relief, many of the major craters have now faded from view, and most of the great walled plains in the central southern regions are barely visible. Other features, such as those in eastern Mare Imbrium, and many of the sharper, fresher-looking craters, remain clearly discernible, their rims and central elevations reflecting the light of the high Sun. All the nearside maria and their associated bays and lakes are plain to see, and they display a surprising range of tone with subtle hints of coloration.

The crater Tycho, in the southern uplands, is surrounded by the most prominent system of rays on the Moon. Tycho is surrounded by a dark collar about 25 km wide, and the rays seem to emanate from this collar rather than the crater itself. A brilliant double ray at 11 o'clock traverses the highlands and crosses south-western Mare Nubium. At 2 o'clock, a sheet of brilliant ray material about 180 km from Tycho is the brightest concentration of ray material on the Moon. Thin rays can be traced for nearly 1300 km across the highlands, almost up to Mare Serenitatis and Mare Fecunditatis, but the prominent ray more than 200 km long that bisects Mare Serenitatis actually originates from Menelaus in the southern mountain border – not from Tycho, as many cursory observers imagine. Tycho's longest ray, more than 1400 km long, reaches across to Mare Nectaris, and traces of ejecta can be made out on the mare itself. From 8 o'clock, a straight ray almost touches the south-western limb.

Through binoculars, the rays surrounding Copernicus are resolved into a highly complex system which extends for more than 800 km over the surrounding region, including eastern Oceanus Procellarum, southern Mare Imbrium and the whole of Mare Insularum. Copernicus' rays are noticeably less bright than those of Tycho, but the largely dark marial background allows finer detail to be observed within them. In places, Kepler's much more compact but equally bright ray system mingles with the ejecta from

Copernicus. In northern Oceanus Procellarum, the brilliant crater Aristarchus is surrounded by a small spray of rays. The trio of Aristarchus, Kepler and Copernicus, along with Tycho in the south, can be seen with the average unaided eye at full Moon. Of the more unusual smaller ray systems, those of Messier in Mare Fecunditatis, Proclus to the west of Mare Crisium, and Thales near the north-eastern limb are worth perusing. Crowning the lunar disk near the northern limb, Anaxagoras has a brilliant spread of rays that stretches across the highlands for more than 400 km. After full Moon, the morning terminator moves round to the farside of the Moon, and the evening terminator begins its trek across the nearside from the east.

▲ Full Moon, imaged on on 29 January 2002 by Mike Goodall, using a Fuji 2600 digicam and an 80 mm refractor.

Mare Orientale, on the south-western limb, is elusive since most of it lies on the farside. It was named Mare Orientale because at the beginning of the 20th century, when it was named, the classical convention for defining east and west on the Moon was still being adhered to. In 1961 the International Astronomical Union adopted the so-called astronautical convention, with the result that the 'Eastern Sea' now lies in the west. Mare Orientale itself is only 300 km across, but occupies the centre of a multiringed basin 900 km in diameter. Lacus Veris and Lacus Autumni lie comfortably on the nearside, and can often be glimpsed easily enough, even when Mare Orientale itself is not visible. Depending on the state of the Moon's libration, any time after full Moon will allow a nice, contrasty, shadow-free view of Mare Orientale (when it is on the Earth-facing hemisphere) and its surrounding lava lakes.

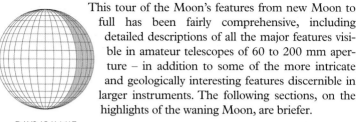

This tour of the Moon's features from new Moon to full has been fairly comprehensive, including detailed descriptions of all the major features visible in amateur telescopes of 60 to 200 mm aperture – in addition to some of the more intricate and geologically interesting features discernible in larger instruments. The following sections, on the highlights of the waning Moon, are briefer.

DAYS 15/16/17

Day fifteen

Essentially a full Moon to the untrained eye, binoculars will show narrow shadows along the evening terminator which indicate the presence of foreshortened features near the limb. Mare Humboldtianum makes a substantial shallow dent in the north-east, along with Gauss, some distance to its south. In the east, the low mountain borders of Mare Marginis, Mare Undarum, Mare Spumans and Mare Smythii

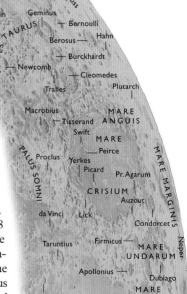

trace thin black outlines around the seas' western edges, and the mass of lava lakes and shallow, dark-floored craters that make up Mare Australe are fascinating to observe at medium to high magnifications. Near the limb, south of Mare Marginis, can be seen Neper (137 km), a very large, well-formed crater with considerable internal terracing and a large central mountain. Other features of interest farther south include the walled plain Kästner (105 km), the clearly defined La Pérouse (78 km) with its prominent bench-like terracing and central peak, smooth-floored Ansgarius (94 km), and the conjoined walled plains of Hecataeus (127 km), Phillips (124 km) and Humboldt in the south-east.

Day sixteen

Evening illumination is beginning to make many of the familiar eastern features increasingly prominent as they take on internal shadow and their eastern internal heights brighten. Mare Crisium's high eastern mountain border displays immense complexity at the beginning of day sixteen. A medium-to-high-power view of Condorcet (74 km), a prominent crater south-east of Promontorium Agarum in Mare Crisium, will show a distinct dark cross on its flat grey floor. Now is the ideal time for high-magnification studies of the inner features and eastern walls of the great line of imposing south-eastern craters that includes Langrenus, Vendelinus, Petavius and Furnerius. These features are covered by the evening terminator by the end of the sixteenth day.

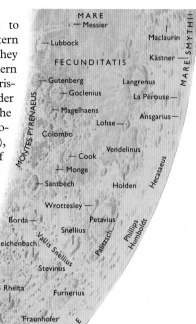

Day seventeen

With the unaided eye, the dent in the terminator caused by a bisected Mare Crisium is easily visible. Early on day seventeen, shadow hides the eastern floor of Mare Crisium and some of Mare Fecunditatis. By the end of the day, the terminator has completely covered Crisium and has immersed half of Fecunditatis. Binoculars clearly show the trio of Endymion, Atlas and Hercules in the north-east, and the line of large craters north of Mare Crisium that includes Cleomedes and Geminus. The southern uplands are beginning to re-establish themselves as a complex mass of prominent craters, but positive identification of most of the medium-sized features in this region is possible only with a telescope at medium to high magnification – plus a good map, of course. Of the more obvious southern features that come into view during the seventeenth day are Vallis Rheita, the crater Janssen, and the prominent cluster of large craters in the Hommel region.

Day eighteen

In the north, Mare Frigoris' eastern reaches dip into the evening terminator. The prominent uplands of Montes Taurus, adjoining Mare Serenitatis' south-eastern border, begin to break into a scattered mass of prominent peaks, and the plains of eastern Mare Tranquillitatis are darkening, revealing their intricate wrinkle ridges. Mare Fecunditatis is now immersed in the lunar night, while Mare Nectaris makes a prominent feature. Its larger outer basin arc of Rupes Altai casts a narrow but increasingly prominent black shadow line from Piccolomini around to the west of Catharina. Janssen is the largest crater crossed by the terminator of day eighteen. Sunset is

DAYS 18/19/20

the best time to view its magnificently sculpted interior and the single prominent rille that curves across its southern floor.

Day nineteen

As eastern Mare Serenitatis begins to vanish from view into the cold lunar night, half of Mare Tranquillitatis is blanketed by the evening terminator, while little Mare Nectaris is completely covered with shadow. Lacus Mortis lurks among the evening shadows in the north, the crater Bürg prominent on its floor, overlooking its rilles. In contrast to these rilles, created by tension in the lunar crust, the large, snake-like wrinkle ridges of eastern Mare

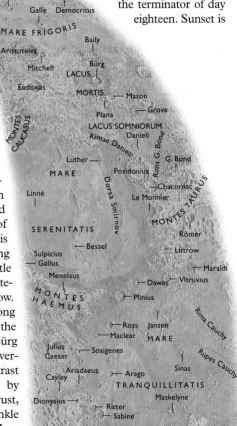

Serenitatis are the results of compressive forces. But the star attraction of the northern terminator during the early part of day nineteen is Posidonius, with its complex floor structure. Less than ten hours after the evening terminator has eaten Posidonius, sunset takes place over the mighty Theophilus chain in the south. Nearby, Rupes Altai now casts a broad black shadow on to the plains to its east – an easy target for binoculars, and a wonderful sight at medium magnification through a telescope if the field of view is wide enough also to encompass the Theophilus chain. The wrinkle ridges in and around Lamont in Mare Tranquillitatis come into prominence at the end of the nineteenth day – a superb target at high magnification in a 150 mm aperture.

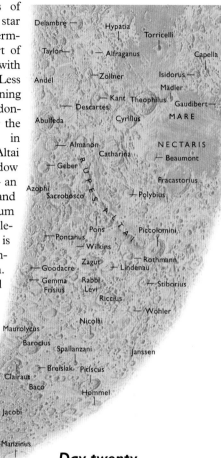

Day twenty

The half-illuminated Mare Serenitatis makes a semicircular impression on the northern terminator, easily visible with the unaided eye, as the last vestiges of the western plains of Mare Tranquillitatis disappear into the lunar night. Binoculars clearly show the prominent duo of Aristoteles and Eudoxus near the evening terminator in the north, in addition to a mass of craters in the southern uplands. The most prominent southern terminator features of day twenty include Abulfeda, Sacrobosco and Maurolycus. The huge spine of Montes Caucasus is beginning to attain prominence towards the end of the twentieth day, and shadows spread across the less prominent Montes Haemus, the evening illumination bringing to light all of the magnificent north-west/south-east sculpture in the vicinity.

Day twenty-one

The waning gibbous phase of day twenty-one is one of the lunation's most spectacular, as the terminator presents such a wide diversity of easily observable features. Mare Serenitatis is no longer on view, half of Mare Frigoris is now cloaked in shadow, and the eastern plains of Mare Imbrium are experiencing an ever-decreasing angle of illumination. Mare Vaporum may be glimpsed just above the centre of the terminator line with a keen unaided eye. A high-magnification telescopic view of sunset over Montes Alpes and the grand Vallis Alpes is a memorable sight. Plato to the west is beginning to fill with shadow, and the isolated mountain ranges and peaks in northern Mare Imbrium start to extend shadowy fingers eastwards. Brilliant on the terminator, the remaining high peaks of Montes Caucasus catch the sunlight past the terminator. In eastern Mare Imbrium, Archimedes, Aristillus, Autolycus and Montes Spitzbergen make an enthralling spectacle, but the broad arc of Montes Apenninus to their south eventually coaxes the eye down towards Eratosthenes and brilliant Copernicus, which is still illuminated by the afternoon Sun. At high magnification with a 150 mm telescope, the Sinus Medii region can be studied in detail – look for the

DAYS 21/22

many intricate rilles, including Rima Ariadaeus, Rima Hyginus and Rimae Triesnecker. Binoculars easily show the major walled plains of Hipparchus and Albategnius near the terminator, while farther west lies the magnificent chain of plains from Ptolemaeus to Arzachel, whose interiors are beginning to fill with evening shadow.

Mösting — Réaumur
Flammarion
Horrocks
Spörer
Fra Mauro
Herschel —
Hipparchus
— Parry
Palisa
Bonpland
Ptolemaeus
Halley
Tolansky
Davy
— Guericke
Albategnius
Klein —
Lassell
Alphonsus
Opelt
Parrot
— Vogel
Alpetragius —
Argelander —
M A R E
Arzachel
Airy
— Gould
Delaunay — Donati
N U B I U M
— Faye
— Wolf
— Thebit
La Caille
Birt
Purbach
Blanchinus
Hesiodus
Apianus
Werner
Regiomontanus
Pitatus
Hell
Aliacensis
Gauricus
Deslandres
— Cichus
Walter
Wurzelbauer
Lexell
Nonius
Huggins Miller
Orontius —
Stöfler
Tycho
Nasireddin
Saussure
Licetus
Heraclitus
Maginus
Porter
Lilius
Clavius
Deluc
— Rutherfurd
Gruemberger
Curtius
Moretus
Short
Rupes Recta

Day twenty-two

Through binoculars, the faint blue tint of earthshine can now be discerned on the unilluminated hemisphere of the last quarter Moon. Plato approaches the evening terminator. Irregularities along its western rim give rise to at least six distinct shadow peaks, projecting across its smooth dark floor. A prominent semicircle of shadow can easily be seen cutting though Plato's western ramparts, marking the edge of a large crustal block that has broken away and slumped into the crater. Shadows cast by Montes Teneriffe, Montes Recti and other peaks point eastwards across the northern Imbrium plains, creating the illusion that the peaks casting them are incredibly high, jagged spires. Eratosthenes, at the western end of Montes Apenninus, is one of the more prominent near-terminator craters in the north. Mare Nubium's eastern edge is slowly introduced to the chill of lunar night early on the twenty-second day. Here, Rupes Recta's brilliant scarp face can be seen illuminated by the low evening Sun. The vast eroded plain of Deslandres broods over the highland terminator to the south, while the interiors of Tycho and Clavius are gradually being shaded in as the Moon assumes a broad crescent phase towards the end of the day.

Day twenty-three

DAYS 23/24/25

Earthshine is now faintly visible with the naked eye. Bisected by the evening terminator, Mare Imbrium in the north and Montes Jura stand out. Promontorium Laplace, with its broad shadow, is an outstanding feature as the terminator approaches it towards the end of day twenty-three. Montes Carpatus and Copernicus are at their best tonight – the play of light from low in the west makes the mighty impact crater and its radial external ridges and furrows appear more impressive than under morning illumination. Bullialdus and its environs in south-western Mare Nubium are fascinating at a

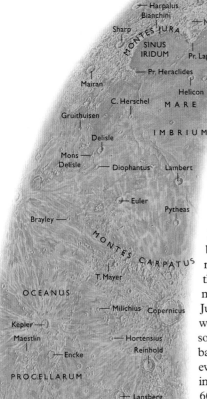

high magnification, and the nearby Rimae Hippalus show up as a set of bold black lines on the eastern edge of Mare Humorum – easy features to resolve through a 60 mm telescope.

Day twenty-four

Mare Imbrium is largely in shadow, and Mare Nubium has vanished into the lunar night. Towards the end of day twenty-four, Mare Humorum is half in darkness, and the naked eye will see it as a dent in the terminator. Montes Jura lie on the terminator, beyond which high peaks at the range's eastern end peek out into the sunlight. Eagle-eyed observers may be able to make out Montes Jura today without optical aid. The wrinkle ridges marking the buried southern wall of the Sinus Iridum basin are more prominent in the evening than they were in the morning, and are an easy target for a 60 mm telescope.

The image labels on the map (left):

Flamsteed

Euclides

Montes Riphaeus

MARE COGNITUM

Letronne

Lubiniezky

Gassendi

Agatharchides

Bullialdus

MARE NUBIUM

MARE HUMORUM

Loewy — Hippalus — König

Rimae Hippalus

Puiseux

Kies

Campanus

Mercator

Vitello

PALUS EPIDEMIARUM

Lee

Capuanus

Hainzel

Epimenides

Mee

Wilhelm

Longomontanus

Schiller — Bayer

Scheiner

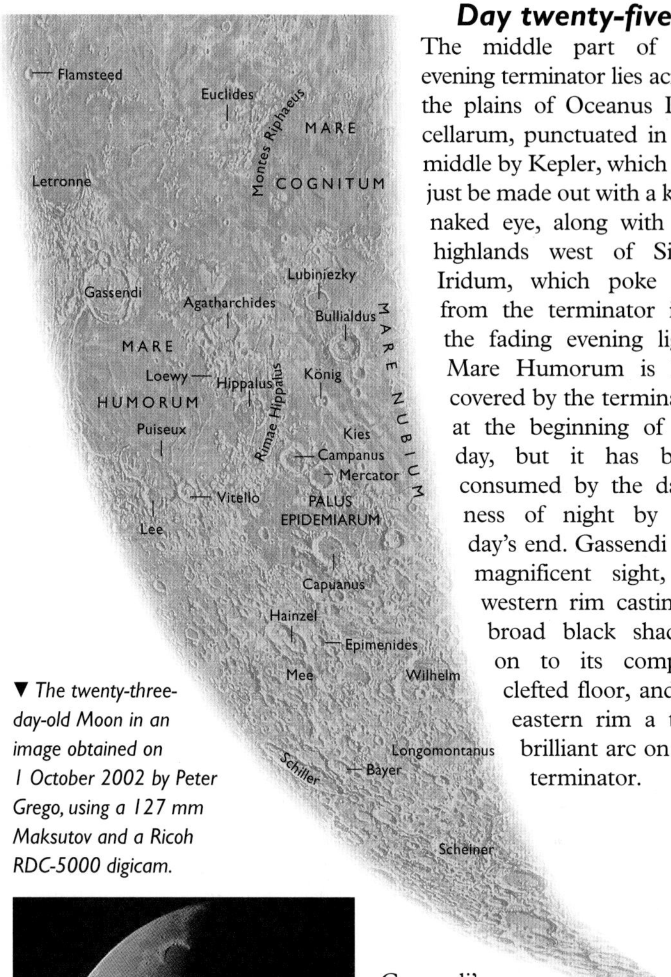

▼ The twenty-three-day-old Moon in an image obtained on 1 October 2002 by Peter Grego, using a 127 mm Maksutov and a Ricoh RDC-5000 digicam.

Day twenty-five

The middle part of the evening terminator lies across the plains of Oceanus Procellarum, punctuated in the middle by Kepler, which can just be made out with a keen naked eye, along with the highlands west of Sinus Iridum, which poke out from the terminator into the fading evening light. Mare Humorum is half covered by the terminator at the beginning of the day, but it has been consumed by the darkness of night by the day's end. Gassendi is a magnificent sight, its western rim casting a broad black shadow on to its complex clefted floor, and its eastern rim a thin brilliant arc on the terminator.

Gassendi's eastern floor appears to dip down beneath the mean level of the rest of the floor, producing a distinct darkening down to the base of the inner eastern wall. In the far south, elongated Schiller and the Schiller Annular Plain are near the terminator, the inner southern ring of the plain appearing rather prominent and casting a broad eastward shadow.

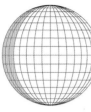

DAYS 26/27

Day twenty-six

When viewed high in a reasonably dark pre-dawn sky, earthshine is a prominent aspect on day twenty-six. Only two major areas are visible without optical aid – the broad dark plain of Oceanus Procellarum, taking up the northern part of the crescent, and the brighter highlands to its south. The Aristarchus region is on the evening terminator, and it is now possible to spot the crater with the naked eye. Under a low evening illumination, a great many of the craters scattered around Oceanus Procellarum seem to lack external relief – they appear to be deep black pits sunken into the mare surface without an appreciably raised rim. This is largely an illusion of lighting combined with the low angle at which these near-limb craters are viewed – the observer

Pythagoras

Babbage

South

SINUS RORIS

Harding

Mons Rümker

Gerard

Naumann

Lichtenberg

Russell

Briggs

Vallis Schröteri

Struve

Eddington

Aristarchus

Schiaparelli

Seleucus

Krafft

OCEANUS

Marius

Cardanus

Galilaei

Olbers

Reiner Gamma

Reiner

Cavalerius

PROCELLARUM

Hedin

Hevelius

is peering directly at an entirely shadowed internal western wall; the bright outer western ramparts and the inner eastern walls are turned away from the observer's line of sight, and are very much foreshortened. However, the evening lighting benefits the appearance of the dome fields west of Marius, the dark side of each raised component being presented nicely for the observer. Of the most prominent southern terminator features, the vast walled plain Schickard and the nearby craters Phocylides, Nasmyth and Wargentin will now repay study.

Day twenty-seven

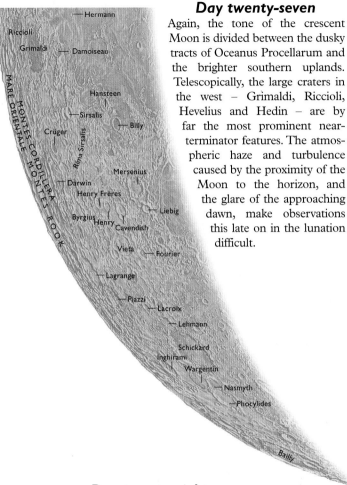

Again, the tone of the crescent Moon is divided between the dusky tracts of Oceanus Procellarum and the brighter southern uplands. Telescopically, the large craters in the west – Grimaldi, Riccioli, Hevelius and Hedin – are by far the most prominent near-terminator features. The atmospheric haze and turbulence caused by the proximity of the Moon to the horizon, and the glare of the approaching dawn, make observations this late on in the lunation difficult.

Day twenty-eight

Approaching the end of the lunation – which actually lasts for 29 days, 12 hours and 44 minutes – it is possible to observe the slender twenty-eight-day-old crescent Moon with the unaided eye (certainly through binoculars) early on day twenty-eight, when the Moon is around 19° west of the Sun. From the northern hemisphere, attempts to view the narrow waning crescent are best made during the autumn (September and October) when the morning ecliptic makes the steepest angle to the horizon, and the twenty-eight-day-old Moon is at its highest above the horizon before sunrise. The best time to observe the waning crescent Moon in the southern hemisphere is during March and April.

— RECORDING YOUR OBSERVATIONS —

With its dark lava plains, jagged mountains and profusion of craters, the Moon is such a beautiful object that the urge to record it accurately has been the aim of observers ever since Thomas Harriot sketched the Moon at the eyepiece of his little refractor in 1609. Until the advent of photography in the mid-19th century, drawing the Moon at the eyepiece was the only way to capture the splendour of the lunar landscape. The arrival of CCD cameras and their increasing availability to amateur astronomers since the 1980s has enabled lunar imaging to take its next great step – high-resolution imaging that captures much of the fine detail that the eye can see.

Conventional photography

The Moon is such a large, bright object that any traditional camera pointed through a telescope can capture a lunar image on photographic film. With luck and experimentation, pleasing images can be obtained by holding a camera close to the eyepiece of an undriven telescope and clicking away. But this is very much a hit-and-miss approach – expensive, too, if out of a whole roll of film you end up with just one or two reasonably good shots.

There are a number of important guidelines that all successful conventional lunar photographers follow, but there is also a knack to it, – a mixture of experience and intuition that cannot be explained clearly in writing. What can be said, though, is that for the very best results a sturdy, equatorially mounted, driven telescope is required, since even with exposures of fractions of a second, the Moon's drift through the field of view of an undriven instrument will cause some image blurring. To obtain the sharpest image possible, the camera should be secured firmly to the telescope so that it is in good alignment with the main light path through the telescope (called its optical axis).

Eyepiece projection photography with a fixed-lens camera

This method of photographing the Moon (also known as afocal photography) came into widespread use in the 1950s when fixed-lens 35 mm cameras became popular. Because the camera lens cannot be removed, the image must be focused on to the film through the telescope eyepiece. This is done by first focusing the image at the eyepiece and then attaching the camera to the telescope eyepiece with the camera focus set at infinity.

The techniques of afocal photography can be used with both conventional and digital fixed-lens cameras alike. However, the simplest and cheapest types of film camera have no means of focusing the image – they have a set focus range from several metres to infinity – nor can the exposure time be adjusted. In spite of their severe limitations, such basic cameras can capture a pleasing image of the Moon. The photographer will usually have to fashion a makeshift adapter for the camera to fit on to the telescope, since the cheapest cameras are not threaded to accept adapters or attachments. All amateur astronomers ought to have in their toolbox some reusable adhesive, such as Blu-Tack and vinyl adhesive tape, suitable for temporarily fixing a lightweight camera to a telescope.

Even though an eyepiece can give a large apparent field of view when viewed with the eye, vignetting – where the image of the Moon is surrounded by a dark circular border – can spoil a photograph taken by this method. The degree of vignetting depends on the size of the camera's fixed lens in relation to the eyepiece lens, on the type of eyepiece used and on the distance of the camera lens from the eyepiece. Vignetting usually happens because the camera lens is much larger than the eyepiece lens. In addition, the eye relief of an eyepiece affects the degree of vignetting: eyepieces with very short eye relief (usually ones of short focal length which give high magnification) are difficult to use afocally with a fixed-lens camera because it is necessary to position the camera so that its lens and the eyepiece lens are as close as possible – so close that they are almost touching. The result is usually highly unsatisfactory, with only a small central image surrounded by a dark border, like viewing through a long tunnel. Eyepieces that afford small apparent fields of view also produce vignetting, so basic types like the Huygenian are to be avoided.

Finally, make notes of your experiments! There is nothing more frustrating than running off a roll or two of lunar exposures, discovering several lunar images that appear to be perfect, only to scratch one's head and wonder exactly which eyepiece was used and what the exposure was.

Advanced lunar photography The most popular type of camera with lunar photographers is the 35 mm single lens reflex (SLR), a design that has been in common use since the 1960s. Some of the older and lower-end SLRs are very basic in design, with exclusively

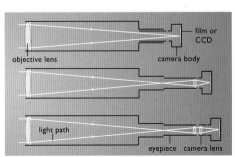

▶ *The three methods of lunar photography at the telescope. Prime focus (top), eyepiece projection (middle) and afocal photography (bottom).*

139

manual controls. The more expensive SLRs are fully electronic and allow the photographer to control every aspect of the camera's functions. A basic SLR is all that is required for lunar photography, and it can deliver just as good a lunar image as can a camera ten times its price. The image entering the SLR is directed by an internal mirror into the viewfinder, allowing a direct view of the subject being photographed. When the shutter is released the mirror flips up, allowing light to fall directly on to the film. The mechanical vibrations introduced by 'shutter slap' can affect the quality of the image, and the mirrors of some cameras can be locked in the 'up' position before the image is taken. To minimize vibration caused by pressing the shutter button, which might blur an image, lunar photographers use a cable release – a function that can be controlled electronically on high-end SLRs.

Telephoto lens photography Telephoto lunar photography has several advantages. The equipment is relatively lightweight and portable, which means that it can be quickly set up in locations where a bulky telescope and its mount could not be. Used with fast film (with a higher ISO rating, hence shorter exposure times), the relatively large field of view (2.5° across with an 800 mm telephoto lens) will not produce a great deal of image drift or blurring. Provided the camera is stably supported (preferably on a tripod), very nice images of the whole Moon, showing detail along the lunar terminator, can be obtained with a 35 mm SLR using a telephoto lens with a focal length of 800 mm to 2000 mm. To find the size of the lunar image on film, simply divide the focal length of the lens (in millimetres) by 110. A 50 mm lens will give a tiny lunar image size, of less than half a millimetre; an 800 mm lens will give an image

◀ *The author's 100 mm refractor set up for prime focus astrophotography with a DSLR camera.*

▲ Prime focus lunar photography can be performed with long-focal-length telephoto lenses or through the telescope. The prime focus image of the Moon on film increases in size with the focal length of the telescope/telephoto lens. Shown here are the comparable sizes on film of prime focus images taken with (from left) 500 mm, 1000 mm and 2000 mm focal-length instruments.

more around 7 mm across, and a 2000 mm lens will produce an image measuring 18 mm. A 2× teleconverter will effectively double the focal length of any telephoto lens, extending a 500 mm focal length to 1000 mm, though the resulting image will be slightly dimmer and inferior to one taken with a 1000 mm telephoto lens. A teleconverter produces a much higher-quality result than you can get by photographically enlarging a smaller image.

Prime focus photography When the camera body (minus the lens) is attached to the telescope with a special adapter, the telescope effectively becomes a large telephoto lens. A telescope with a focal length of less than 2000 mm can project an image of the entire Moon on to a single 35 mm frame. This method of prime focus photography (so called because the image falling on to the film or CCD chip is at the telescope's prime focus, the light being beamed straight from the primary lens) is the best way to capture a sequence of lunar phases. It is quite easy to focus the Moon through the camera's viewfinder, and exposure times must be judged according to the focal ratio of the telescope, the ISO rating of the film and the Moon's phase (see below).

Eyepiece projection Prime focus photography is ideal for capturing the entire lunar disk, but for closer, higher-magnification views the telescope's eyepiece can be used to project the image directly on to the film. Eyepiece projection adapters are available which fit on to the telescope and hold the camera firmly near the eyepiece. A standard Plössl eyepiece delivers nice crisp images with a flat field of view. Experiment with different focal lengths, and bracket the exposure times (in other words, take a series of photographs covering a range of exposure times) to find the combination that suits your requirements. Remember to note the results of your experiments for future reference.

EXPOSURE TIME (SECONDS) FOR LUNAR PHASES			
Lunar phase	1° frame field	0.5° field (whole Moon)	0.25° field (with terminator)
Thin crescent	1/125	1/30	1
Broad crescent	1/250	1/60	1/2
Quarter phase	1/500	1/125	1/4
Gibbous	1/1000	1/250	1/8
Full	1/2000	1/1000	1/30

Film types and exposure time For the beginning lunar photographer, standard ISO 200 colour film is ideal to experiment with since it is cheap and widely available, though the quality will vary from brand to brand (and even in different batches of the same cheap brand). The ISO rating of a film denotes its speed – the higher the ISO number, the faster the film and the shorter the exposure time required. The Moon is so bright that it is possible to use very slow film to capture it – even as slow as ISO 25. Slower films are less grainy and so can allow greater enlargement than faster films without the grain showing through.

Formulae for calculating lunar exposures take into account the aperture and focal length of the lens/telescope, the camera *f*/stop employed, the film speed and the Moon's phase. The brightness of the Moon and hence the exposure time required are also affected by the Moon's altitude and the presence of cloud. The above table is based on my experience with a 100 mm Maksutov (*f*/10), a basic Zenit SLR camera and ISO 200 Kodachrome transparency film.

Digital imaging

Digital cameras With most digital cameras the lens cannot be removed, so the principles of photographing the Moon through them are the same as for fixed-lens conventional film cameras and the afocal method. Digital cameras are versatile, with a variety of functions that can be adjusted to suit the conditions, and images can be instantly reviewed on the camera's LCD screen to determine whether they are worth keeping. After manipulation with an image processing program, the end results, both on screen and in print, can be superb.

A digital camera's resolving power depends on the size of its CCD (charged coupled device) – a tiny electrical chip with an array of minute photosensitive squares (pixels) immediately behind the camera lens. Cameras with the highest pixel rating will capture the highest resolution single-shot images of the Moon. Digicams can store images in a range of resolutions – the lower the resolution, the smaller the image file and the more images can be stored in the camera. Digicams of 3 megapixels or higher can capture impressive single-shot views of the Moon, while images secured with high-end 8 megapixel (or higher) digicams can produce crisp images capable of being enlarged. Many digicams are capable of taking video clips which can be computer processed to produce high-resolution still images, overcoming the effects of poor seeing.

Focusing can be a problem with digicams, as their TFT screens are usually quite small, some measuring less than 4 cm (diagonally). First, focus the image with your eye through the telescope eyepiece and then attach the digicam to the eyepiece; use zoom to zero-in on the lunar terminator, where fine focusing will prove easiest. If your camera can be hooked up to a TV monitor, the larger image will be far easier to focus by.

Vignetting can be avoided by using the zoom facility. If the edge of the field appears dark or hazy, an adjustment in zoom can remove vignetting altogether. A high degree of zoom brings into play the camera's 'digital zoom' and should be avoided, as in this mode the image is usually of low resolution. It is better to enlarge a high-resolution image after it has been captured than to zoom for a close-up (low-resolution) image in the first place.

Gauging exposure can prove troublesome. The camera's built-in autoexposure meter may overexpose the Moon, particularly if the whole Moon is being imaged and is not centred in the field of view. Autoexposure usually works best if there is a uniformly bright field, so your digicam will probably judge exposures well with lunar close-ups. Many digicams have exposure variation capabilities equivalent to two *f*/stops above and below the optimum, so use this if you are having problems with overexposure or underexposure. The type of light conditions selected can affect exposure too, so experiment with various combinations and note which ones appear to work best.

Digicams can introduce vivid colours that bear no resemblance to the view through the eyepiece. Many digital photographers mute these

▶ *The author's Ricoh RDC-5000 digicam set up with his 80 mm achromatic refractor for afocal photography through a 32 mm Plössl eyepiece.*

tones in the processing, or decide to convert the images to black-and-white, which can appear much more pleasing to the eye than a multi-coloured Moon. If your camera has the ability to photograph in black-and-white, give it a go – black-and-white images can appear much sharper than colour images, and the smaller image files take up less space in your camera's memory.

Camcorders Although the Moon is not a dynamic world, its shadows creeping at a snail's pace across the surface near the terminator, video footage of the Moon is fascinating to watch, and conveys a sense of being at the eyepiece far more than a still picture can. Video captures the atmospheric shimmer as it momentarily distorts the view, and a sweep at high magnification along the terminator using the telescope's slow motion controls captures the experience of observing. Camcorders are also ideal for recording lunar eclipses and planetary occultations (see Chapter 6).

Camcorders are heavier than digicams, and the coupling between the camcorder and the eyepiece must be rigid. Adapters are available that slot into the eyepiece socket and hold the camcorder on an adjustable arm so that it can be brought up close to the telescope eyepiece and locked into position. Digital camcorders are the lightest and most versatile available, and their images can be edited on the computer.

Camcorder imaging uses the same principles as conventional or

◀ The author's 150 mm f/8 achromatic refractor with digital camcorder set up with a zoom eyepiece for afocal video photography.

digicam afocal photography. The problems of eyepiece selection and vignetting are overcome as with digital photography (see above). It is possible to put together a wonderful tour of the Moon and its terminator, taking time to zoom in on interesting features – a fine presentation at any astronomical society meeting. By downloading digital footage on to your computer, images can be grabbed individually (at low resolution) or digitally stacked to produce detailed, high-resolution images. The same method is used to produce high-resolution webcam images (explained below). It is essential, when embarking upon digital video editing, to have at least 5 GB free on your computer's hard drive. The resulting edited videos can be transferred to CD-ROM or DVD-ROM and then removed from your hard drive to free up more space.

Webcams The use of webcams to capture high-resolution lunar and planetary images has been growing in popularity since the early 1990s. Webcams are only a fraction of the cost of dedicated astronomical CCD cameras, and they are lightweight and versatile. Many experienced lunar and planetary imagers prefer webcams because their high frame rate makes focusing easy and allows more images to be captured in a given time. Any off-the-shelf webcam attached to a computer at one end and a telescope at the other can be used to image the Moon and the brighter planets.

A webcam is usually positioned at the telescope's prime focus, in place of the eyepiece. Several models have removable lenses into which can be screwed commercially available adapters. CCDs are highly sensitive to infrared (IR) radiation, and the lens assembly contains an IR blocking filter. Without an IR filter, a perfectly clean focus through a refractor is not possible because IR radiation is focused differently from visible light. However, there are available IR blocking filters that are placed between telescope and webcam, allowing only visible light to pass through to a sharp focus.

Although webcams are used at the prime focus, the image of the Moon they produce is at quite a substantial magnification, and only the shortest-focal-length instruments will allow the whole Moon to be viewed in the field. The CCD chip itself is quite small, and only part of the image is projected on to it. For example, my short-focal-length (f/5) 80 mm refractor requires the use of a focal reducer (a device that decreases the magnification) in order to fit the entire Moon into the field of view.

Since a webcam lacks a viewfinder or LCD screen, focusing can be a trying experience. There is no point focusing the Moon through the eyepiece and replacing the eyepiece with the webcam, as there will be a substantial difference in where the focus lies. It is best to focus the webcam on the lunar terminator while you have a clear view of the computer monitor, which is best done with a laptop in the field, next to the telescope. Even a slight knock can push the Moon out of the small field of view – this is where a really well-aligned finderscope is extremely useful. If the webcam is connected to a computer indoors, focusing may well entail some rushing to and fro until a good focus is achieved. Once a good focus has been found, carefully mark the position of the focuser tube with a fine-tipped permanent marker to avoid having to go through the whole process again the next time you want to do some imaging. Of course, the focus will need tweaking slightly each time you set up, since a fraction of a millimetre can make all the difference between an acceptable focus and a razor-sharp one. Electric focusers are highly desirable for webcam imaging, and being able to adjust the focus at leisure from indoors at the computer can make a great deal of difference to your enjoyment and the quality of your images.

The webcam manufacturer's own software is used to record image

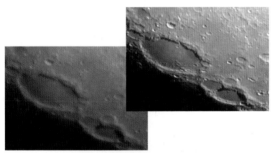

◀ CCD imaging with a webcam can produce startling results from a seemingly mediocre video sequence. Shown here is a comparison between a single unprocessed captured CCD frame at left (a BMP image from the original AVI file) and the end result of stacking and processing 40 of the best such BMP images. The features shown are Schickard, Phocylides and Nasmyth.

◀ The author's Philips ToUcam 740 Pro webcam is used in conjunction with a ×2 Barlow lens at the prime focus of his 150 mm achromatic refractor. Behind can be seen the laptop to which the webcam is linked, for use in the field.

sequences as AVI files from the webcam to the computer. The image download rate will vary with the type of equipment used. Using an older USB-1 camera of 640 × 480 resolution, the rate should be limited to a maximum of 5 frames per second to avoid lossy compression. Much higher rates of up to 30 fps with no compression are possible with USB2-enabled high-end webcams. The image can be adjusted to produce the best balance of exposure, brightness, contrast and colour, but often the Moon can be successfully imaged in full auto mode in short bursts. As with digicam imaging, the use of black and white (greyscale) can simplify matters, especially if you're using a short-focal-length achromat prone to producing a degree of false colour.

Image-editing software, such as the freeware Astrostack, is then used to extract individual images from the AVI. A well-aligned and smoothly operating equatorial mount is essential to good webcam imaging, as a misaligned mount will produce image drift even in short capture sequences – this produces problems when attempting to align and stack the images. AVI files are chopped into individual BMP (bitmap format) frames by the processing software, each of which can be checked for quality either visually (a laborious process) or automatically by the software itself. The best images can be stacked and combined, manually or automatically, to remove noise (improving the signal-to-noise ratio) and eliminate artefacts. The combined stacked image is then processed to bring out detail – unsharp masking being one of the most useful techniques for achieving a clean image. This automated process uses a blurred image of the original in conjunction with the original itself to bring out detail. Unsharp masking can magically transform mediocre images, but beware of applying too much image processing because spurious artefacts can – and do – arise as a result of overzealous manipulation, and fine tonal detail can be obliterated.

Drawing the Moon

Lunar observers of the early 21st century are fully aware that the Moon has been mapped by space probes in exquisite detail. So, what possible reason is there to spend hours making drawings of lunar features? On the face of it the activity belongs to the distant past – and so it does, in a sense, as no lunar observer actually believes that there remains a significant Moon-mapping role left for the amateur. But there is no doubt that lunar observers who take the time and trouble to make accurate lunar drawings will discover an immensely useful and rewarding activity which improves every single aspect of their observing skills.

The observer's ability to discern fine detail constantly improves as more hours are spent at the eyepiece. By concentrating on drawing skills, the observer learns to attend to detail instead of allowing the eye to wander on to the more obvious features. The discipline of accurate lunar drawing will pay off in other fields of amateur astronomy which require pencil work, in particular planetary and deep-sky observation.

During an apprenticeship in lunar drawing, the apparent confusion of the Moon's landscape, with its seemingly arcane nomenclature, becomes increasingly familiar. Over the years I have seen the work of hundreds of newcomers to lunar observation in the Society for Popular Astronomy's Lunar Section, and without exception everyone's observing abilities and knowledge of the Moon have been enhanced. I have been through the apprenticeship myself, starting off with a small refractor, and I continue to learn each time magnified moonlight impacts on my retina.

It is important to be confident in your drawing ability – so you have my permission to disregard everything your art teacher ever told you if your artistic experience at school was not a positive one. The lunar observer is not some kind of nocturnal art student, seeking good marks for artistic flair or aesthetic appeal. Observational honesty and accuracy counts above all. Most Moon drawings seen in books and magazines are the product of skilful observing and many hours of practice.

Pencil sketches Invest in a set of soft-leaded pencils from HB to 5B and an A5 (or $5\frac{1}{2} \times 8\frac{1}{2}$ inch) pad of smooth art paper. If your sketching skills need to be polished, there is no better way than to start by copying sections of lunar photographs which appear in books and magazines. First, lightly sketch the basic outlines with a soft pencil – this gives you the chance to erase any mistakes. When shading dark areas, try to apply minimal pressure to the paper. The darkest areas are

▲ By using a low-resolution, low-contrast print of a digicam image as a template for an observational drawing, general positional accuracy can be achieved, allowing time to concentrate on fine detail. Here, the author's digicam image is compared with his actual (unretouched) telescopic drawing of Janssen, both made on 23 March 2000 with a 250 mm reflector.

▲ *How to make a pencil drawing of the Moon at the telescope eyepiece. Basic outlines are sketched in lightly at first, taking care to position the larger features accurately in relation with one another.*

Shadows are then implanted, followed by fine detail and tone. This drawing is based on an observation of the craters Steinheil and Watt, made with a 100 mm Maksutov.

ideally shaded in layers, and not in one mad burst. After several sessions of leisurely 'armchair' Moon drawing you will surprise yourself at how quickly you see improvement. The most important thing to remember is to be patient. Do not rush, even if you are only practising.

Out in the field, when the real Moon is sharply focused in the eyepiece, do not be intimidated by the wealth of detail visible. First, it is important to orientate yourself. If you are not sure what features you are looking at, use a good map of the Moon to get your bearings. Your eye will be drawn to the Moon's terminator, where most relief detail is visible under the low angle of sunlight. The area you choose to draw should be quite small, such as an individual crater, and close to the terminator so that it has good contrast. If a feature you have chosen to draw is not marked on the map, then make a note of nearby features that can be identified and indicate their positions in relation to your subject.

If possible, go back indoors and (copying from a map or photograph) make a light outline drawing of your chosen area. This should be at least 75 mm across, and larger if you are attempting to portray a region full of fine detail. Completing the first stage of a drawing indoors in this way will save you a lot of time and will allow you to concentrate on filling in detail at the eyepiece instead of worrying about proportion and configuration.

Allow an hour or two for each drawing session. Patience is vital – a rushed sketch is bound to be inaccurate. Describe unusual and interesting aspects in short written notes. It is up to you as the observer to decide how lengthy or detailed these notes need to be. My own preference is to make sure the drawing does most of the talking by making as good a job of it as I can, writing just a few relevant words at the eyepiece to point out any aspects that cannot adequately be conveyed in the sketch. Next to the drawing, note the usual important observing information, such as date, start and finish times (in Universal Time, UT), details of the instrument and magnification, and the seeing. More technical information, such as the position of the terminator and the extent of libration, may be calculated and written into the observing notes back indoors.

Line drawing Features can be depicted in simple line form as an alternative to making shaded drawings. Use bold lines for the most prominent features, like crater rims and the sharp outlines of black lunar shadows. Showing lunar mountains as upturned 'V' shapes may be fine for cartoonists, but not for lunar observers. If, say, an area is full of rough terrain, simply label it on your line drawing as 'rough terrain' – do not attempt to depict it as rough by filling the area with a mass of dots and jagged squiggles. More subtle features, such as low lunar domes, are best recorded with light thin lines. Dashed lines can be used for features such as rays, and dotted lines to mark the boundaries between areas of different tone or coloration.

Line drawings require plenty of descriptive notes, more so than a good tonal pencil drawing. The line drawing method has the advantage of being quick and requires a minimum of drawing ability. When done properly, the result can be as accurate and as full of information as any toned pencil drawing.

Intensity estimations Another lunar sketching technique is to draw the boundaries between areas of different brightness, and then estimate those brightnesses on a scale of 0 to 10. In his book *The Moon* (1896), the 19th-century English amateur Thomas Elger quoted a scale of intensities adopted by the German astronomer

ELGER'S SCALE OF INTENSITIES FOR LUNAR DRAWINGS	
0	Black – for the darkest of lunar shadows
1	Very dark greyish black – dark features under extremely shallow illumination
2	Dark grey – the southern half of Grimaldi's floor
3	Medium grey – the northern half of Grimaldi's floor
4	Yellow grey (subtle) – general tone of area west of Proclus
5	Pure light grey – general tone of Archimedes' floor
6	Light whitish grey – the ray system of Copernicus
7	Greyish white – the ray system of Kepler
8	Pure white – the southern floor of Copernicus
9	Glittering white – Tycho's rim
10	Brilliant white – the bright central peak of Aristarchus

Johann Schröter in the previous century, and this is still the scale used by many lunar observers today. The examples of tones described in the table here are for a general low-power telescopic view (50×). Each area will, of course, show smaller-scale gradations of tone under higher-powered scrutiny.

Copying up Keep your lunar drawings stored safely. Unless it is absolutely necessary, it is not wise to lend your original drawings to others or send them unregistered through the post. Neat copies of your original observational drawings are best made as soon as possible after the observing session while the visual impression is still fresh in your mind. Never discard an original drawing – retain it as the basis for any subsequent copies that need to be made, for example, for the records of any astronomical societies to which you belong, or for publication. Photocopies are not good enough to submit to society observing sections or for publication, but a good laser print or digitally scanned drawing is usually acceptable – some magazines prefer artwork to be submitted on disk or by email.

There are a variety of media in which copied drawings may be made. Soft pencil on smooth art paper is by far the quickest and least fussy form, and is my personal favourite. Ink washes or gouache paintings are suitable for exhibition, but they are more time-consuming to execute than pencil drawings. Whatever the medium, make sure the finished result is sprayed with fixative to avoid smudging and unsightly fingerprints (usually other people's!).

Stippling is a technique that uses minute, closely spaced black ink dots to convey the illusion of shade – the closer and bigger the dot, the more intense the tone. Because stippled drawings are composed of hundreds upon hundreds of black dots, they reproduce well when photocopied. The technique appears very impressive when done expertly, as evidenced in the splendid work of the late Harold Hill and current BAA Lunar Section members Colin Ebdon, Nigel Longshaw and Phil Morgan. A set of technical pens with a range of nib size is essential; attempts to get away with using a normal fountain pen, biro or felt-tip will produce inferior results. A great deal of skill and practice with a steady hand is needed to approach anything like what could be considered a good standard. Unlike pencil, the technique is unforgiving and does not welcome alteration. Several misplaced dots grouped closely together could give the impression of a non-existent feature. When done by anyone less than an expert, stippling looks plain awful.

Whichever technique you use to record the Moon's surface, drawing the Moon is easily the best way to learn your way around our only natural satellite and to get a feel for the lunar landscape in all its majestic forms.

— *ECLIPSES AND OCCULTATIONS* —

Every planet, satellite, asteroid and comet – indeed, every solid object in the Solar System illuminated by the Sun – casts a shadow into space. The dimensions of an object's shadow depend on its size, shape and distance from the Sun. Because the Sun is an extended object, and not a point source, every shadow is made up of two components: the umbra and the penumbra. The umbra is the darkest part of the shadow. From inside the umbra the Sun is totally obscured, and in the depths of space (in most cases) this is a pretty dark environment. An object close to the Sun casts a shorter umbral shadow than a similar sized object far away from the Sun. The penumbra is far less dark than the umbra, and inside it only part of the Sun is obscured by the object. Theoretically, the length of a penumbral shadow is infinite.

The Earth's umbral shadow cone is 1.4 million km long. At the distance of the Moon, a section through the Earth's umbra is more than 9000 km in diameter, and the penumbra is around 17,000 km across. The Moon itself casts an umbral shadow which occasionally – and then just barely – exceeds the Earth–Moon distance.

From time to time the Moon appears to pass directly in front of the Sun, causing a solar eclipse; the Moon occasionally enters the Earth's shadow, and undergoes a lunar eclipse. Because the plane of the Moon's orbit is inclined by about 5° to the ecliptic, these eclipse-producing alignments do not occur every month – that would only happen if the Moon's orbit lay in the plane of the ecliptic. Orbital geometry dictates that the Earth can experience a maximum of either five solar and two lunar eclipses, or four solar and three lunar eclipses, in any single year.

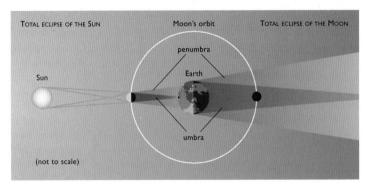

TOTAL ECLIPSE OF THE SUN — Moon's orbit — TOTAL ECLIPSE OF THE MOON

penumbra

Earth

Sun

umbra

(not to scale)

▲ *A solar eclipse occurs when the Moon's shadow falls on the Earth. A lunar eclipse occurs when the Moon passes through Earth's shadow. When the* *Moon passes through the penumbra, a very faint penumbral eclipse is produced; if it goes on to pass through the umbra, a much darker umbral eclipse takes place.*

Solar eclipses

One of the Solar System's greatest coincidences is the fact that the Sun and Moon each have an apparent angular diameter of around half a degree as viewed from the Earth's surface. As a result, the Moon is capable of totally obscuring the bright solar disk for a few minutes during a total eclipse. The Moon's apparent angular diameter actually varies between 33′ 29″ at perigee (when it is nearest the Earth) and 29′ 23″ at apogee (farthest from the Earth). The Sun's apparent angular diameter varies between 32′ 36″ at perihelion (when the Earth is closest to the Sun) and 31′ 32″ at aphelion (when the Earth is farthest from the Sun). Therefore not all eclipses include a period of totality, even when the centre of the Moon passes directly over the centre of the Sun – in the most extreme cases the Moon can be more than 3′ (one-tenth of its apparent diameter) smaller than the Sun, producing an *annular eclipse*. On these occasions, the Moon is surrounded at mid-eclipse by a spectacular brilliant ring of sunlight. At the annular solar eclipse of 1836 the English astronomer Francis Baily was first to describe the so-called Baily's beads, a phenomenon caused by shafts of sunlight shining though valleys on the Moon's limb when the Sun is almost completely covered.

A *total eclipse* of the Sun is one of nature's most awesome sights. For a brief time, sadly never more than 7 minutes and 40 seconds, the solar disk is completely hidden by the Moon. Darkness, coldness and an eerie silence descend upon the observing site. The brighter stars and planets come into view. From the edge of the Moon's dark disk spring out the red prominences of the Sun's chromosphere, and spreading away from it are the pearly streamers of the corona, the outer solar atmosphere.

In addition to the sheer sense of awe engendered around the world by these events, the scientific value of their observation has proved incalculable. Until recently the only prospect of studying the Sun's corona was to await the darkness of totality. Astronomers have literally gone to the ends of the Earth to observe total solar eclipses, taking with them tonnes of equipment and plenty of hope that the weather will be favourable. Thanks to the advent of advanced technology and orbiting solar observatories, serious solar research can now be conducted at all times.

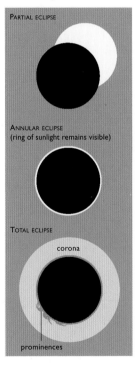

▲ *The various types of solar eclipse. Total eclipses allow the Sun's corona to be observed.*

Extreme care should be taken when observing solar eclipses because the Sun is a million times too bright for direct visual observation. There are many accounts of people's eyesight being permanently damaged after brief flirtations with cheap glass or plastic filters which fit over the telescope eyepiece. It is not only the Sun's visible light, but also its infrared radiation – its heat – that is focused by a telescope. A small spot on the retina of my right eye has been permanently damaged as a result of using an unsuitable solar filter – I would not like anyone to suffer a similar accident. Solar filters that fit over the eyepiece should be outlawed, since they are prone to cracking and melting, allowing a flood of solar radiation to strike the unprotected eye. Full-aperture solar filters made by reputable manufacturers are available, and these provide adequate protection as long as they are undamaged and properly secured over the aperture. By far the best way of viewing the partial phases of a solar eclipse (indeed, the best way of conducting general solar observation) is to project the Sun's image on to a shaded white card. One final word of caution: the objective lens of any finder telescope attached to the main instrument should be capped while observing the Sun. There have been cases of burns to clothing and skin because this simple precaution had been overlooked.

Lunar eclipses

When the Moon traverses the penumbral shadow of the Earth alone, avoiding the umbra, there is a *penumbral eclipse*. Though they are interesting to observe, these events are rather unspectacular, even when the bulk of the lunar disk is immersed in penumbral shadow. For example, the penumbral eclipse of 14 April 1987 was very difficult to detect visually – some observers even doubted the accuracy of the ephemeris predictions. Only the most enthusiastic of amateur astronomers will bother to set their alarm for, say, three in the morning to view this sort of eclipse.

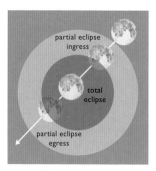

▲ *The progress of a lunar eclipse, as the Moon passes through the penumbra and umbra of the Earth's shadow.*

There's something decidedly eerie about watching the bright full Moon slowly sink into the umbral shadow of the Earth and turn deep orange or red. *Partial umbral eclipses* take place when the Moon passes through the penumbra and then edges partially into the umbra, producing a dark, curved 'bite' out of the edge of the Moon. The magnitude of an eclipse is measured in lunar diameters, from the edge of the umbra. Partial lunar eclipses are assigned a magnitude value from 0 to 1: for example, a magnitude 0.5 partial eclipse will see half of the Moon's diameter covered by the umbra.

▶ The partial phase of the total lunar eclipse of 9 January 2001, as the Moon moves out of the Earth's umbral shadow. This digital photograph was taken afocally by Peter Grego, using a 250 mm Newtonian reflector.

In some eclipses the edge of the umbra fades gradually into the penumbra, yet at other times it can appear much more sharply delineated. The umbra may be noticeably coloured, but if the eclipse is partial and of a small magnitude, the glare of the uneclipsed portion of the Moon will make it difficult to estimate the hues and tones within the umbra. For the same reason, with the naked eye alone it can be difficult to distinguish the eclipsed part of the Moon during the first or last stages of a partial eclipse, and short-exposure photographs of partial lunar eclipses show little detail or colour within the umbra.

The *total lunar eclipse* is one of astronomy's most beautiful sights. At totality the Moon is completely covered by the umbral shadow, but it never vanishes from sight altogether because the Earth's atmosphere refracts sunlight on to the lunar surface. The maximum possible magnitude of a total lunar eclipse is 1.888 (which is to say that the edge of the Moon farthest from the edge of the umbra is separated from the edge of the umbra itself by 1.888 lunar diameters at mid-eclipse), which equates to 1 hour and 42 minutes of totality. No two total lunar eclipses are the same: the hues, colour distribution and intensity of the umbra always vary. Cloud and high-altitude dust in our own atmosphere affects the intensity of the eclipse.

A scale for classifying total lunar eclipses was devised by the French astronomer André Danjon. Danjon's scale (see table page 156) takes into account the brightness and colours of the umbra and serves as a rough guide to the observer. Danjon attempted to find a correlation between solar activity and lunar eclipses, and after examining a long series of accounts he concluded that there was a definite link between solar minimum (when the Sun is least active) and the intensity and redness of the umbra. However, a more definite and predictable relationship exists between terrestrial volcanism and eclipse intensity. The

THE DANJON SCALE OF TOTAL LUNAR ECLIPSE INTENSITY	
0	Very dark. Moon exceedingly dim, especially at mid-totality
1	Dark, grey or brown in colour. Lunar details distinguishable only with difficulty
2	Deep red or rust-coloured. Very dark central shadow, outer edge of umbra relatively bright
3	Brick-red. Umbra has bright or yellow edge
4	Very bright copper-red or orange. Umbra possesses very bright, bluish edge

huge amount of dust released by the eruption of Krakatoa in August 1883 seems to have been responsible for the dark eclipses of October 1884 and September 1888 – and by Danjon's criteria, these eclipses ought to have been bright. The darkness of December 1992's eclipse is thought to have been produced largely by the dust released by the eruption of Mount Pinatubo in the Philippines in June 1991, and this too should have been a bright eclipse according to Danjon's scheme.

Observing lunar eclipses

As with any transient celestial event, successful observation of a lunar eclipse depends on good planning, especially if you intend to make detailed observations or secure a series of images as the eclipse unfolds. Make sure in advance that the Moon will not be hidden behind any buildings or trees at your observing site. It is wise to be aware of weather predictions. Televised weather forecasts usually give little indication of what overnight observing conditions will be like. Fortunately, detailed weather forecasts are available at a variety of internet sites, some of them catering specially for amateur astronomers. Take a timetable of the event with you, plus any other information that may be of use, such as a timetable of any occultations of stars predicted during the eclipse. A notepad and pencil are essential, and you may find it useful to take two flashlights – a bright one for general use, and a small observing one with a red bulb for making notes at the eyepiece.

Naked eye Even without optical aid, the beauty of a lunar eclipse can still be appreciated because of the largeness of the lunar disk. If you happen to be without binoculars or telescope you can still attempt a series of annotated line drawings on blanks prepared beforehand – circles measuring at least 100 mm in diameter, drawn on white paper. Try to make sketches at 15 minute intervals. Record the definition of the umbra's edge and the colours and intensities observed within it – you may be surprised at just how much you can see on the Moon's half-degree diameter disk with the unaided eye.

Using nothing more than a smooth silvery Christmas tree ball or a convex mirror (like a car's rear-view mirror) a good estimate of the brightness of the totally eclipsed Moon can be made. The Moon's small reflected image is compared with a star or planet of known magnitude by holding the reflection of the Moon next to it; this means turning one's back on the Moon, and locating a star or planet on the opposite side of the sky. Using a similar method I estimated the totally eclipsed Moon of 4 April 1996 to be as bright as nearby Eta Boötis (Muphrid), at magnitude 2.7. At the next total lunar eclipse, of 27 September 1996, I estimated the Moon to be slightly brighter than Gamma Pegasi (Algenib), at magnitude 2.8.

Binocular and telescopic eclipse observation Steadily held binoculars give the observer the best overall view of a lunar eclipse. Through binoculars the colours can be striking, and the darkened ruddy lunar globe surrounded by a dark starry field takes on a three-dimensional quality. One popular eclipse activity is to time the passing of the edge of the umbra over certain prominent lunar features to the nearest minute. The recommended features (most of them bright spots) include the craters Aristarchus, Kepler, Copernicus, Tycho, Plato, Manilius and Proclus. A contact timing should be made on the feature's immersion into and emergence from the shadow. Such an exercise can reveal discrepancies between the actual and predicted times of contact. The width and definition of the edge of the umbra should be noted, along with any observed irregularities in its outline.

Lunar eclipse photography Very pleasing results are to be had with simple conventional photographic equipment – indeed, this is one of the few celestial spectacles whose subtle colours and tones record better on photographic film than on digital cameras. An undriven wide-field multiple exposure showing the passage of the Moon through the sky and the progress of the event can be obtained using a simple camera with a time exposure facility. With a telephoto lens attached to an equatorial drive, a triple exposure can be taken showing three side-by-side views of the Moon at immersion, mid-totality and emergence. This type of photograph is stunning and reveals the true extent of the umbral shadow. Single shots of the whole lunar disk in eclipse should be adjusted for exposure according to how much of the Moon is covered with shadow.

Because the brightness of eclipses varies so considerably, no firm guidelines can be given for optimum exposures for totality. A very dark eclipse may well require more than a minute's exposure to reveal detail on the lunar disk, whereas a brighter one may need just a few seconds. It is wise, in any circumstances, to bracket your exposures and experiment with different times as the event unravels. If you are covering the entire eclipse from start to finish, take your pictures at set intervals of, say, 10 minutes – that way you will secure a nice-looking sequence. If you wish to record anything of the subtle colours and tones within the

◄ *The total phase of the total lunar eclipse of 9 January 2001. The photograph was taken by Paul Stephens, using the prime focus photography method, with a 300 mm Newtonian reflector.*

umbra, the partial phases will need a little more exposure than if you were taking photographs of the Moon's regular phases.

On standard ISO 200 film and with the camera set at $f/8$, the uneclipsed full Moon registers in just 1/500s. An exposure of 1/250s is suitable for the early phases of the eclipse, and a half-eclipsed Moon requires 1/125s. A 75% eclipse requires an exposure of 1/60s, but for totality itself the exposure depends on the intensity of the eclipse. It is a good idea to take a variety of exposures in any case, to be sure of getting at least one really good shot.

Lunar occultations

Travelling at an angular velocity of over 13° each day, the Moon ploughs a half-degree wide swathe through the stars. The exact path varies from one orbit to another, but is always contained within a band 5° either side of the ecliptic. A lunar occultation takes place when a star within this band disappears behind the Moon's limb. Because the Moon has no appreciable atmosphere, and most of the stars have negligible apparent diameters, most occultation events happen very suddenly, like switching off a light. Some stars with large apparent diameters may not be extinguished instantaneously, but appear to fade in a fraction of a second. The task of the occultation observer is to record accurately the time of a star's disappearance and/or its reappearance from behind the Moon.

Stellar occultations have been used to determine the Moon's physical shape and its motion with incredible accuracy, and they provided an invaluable parameter for monitoring Terrestrial Time and the Earth's rotation. Stellar occultation timings by amateurs remain vitally important in refining our still-patchy knowledge of the Moon's limb profile, which constantly varies because of libration.

Fadings or staggered effects in occultations can suggest previously unsuspected double or multiple star systems. By an amazing stroke of good fortune, I observed such a phenomenon during the total lunar eclipse of 17 August 1989, where the known double star 44 Capricorni appeared to just skim along the edge of the Moon (an event known as a *grazing occultation*). Over a period of a few minutes the star faded to around ninth magnitude, displayed variations in which it wavered around a magnitude higher, and finally brightened to its usual sixth magnitude as it rapidly scintillated. My observation provided strong evidence for what had been only a suspected multiple star system, perhaps a 'double double' like Epsilon Lyrae. The US Naval Observatory lists 44 Capricorni as a double star whose components' own dual status is uncertain. More advanced investigations will hopefully settle the matter once and for all.

Occultation observers should be equipped with a sturdily mounted and optically good telescope, an eyepiece giving a medium to high magnification, a stopwatch capable of an accuracy of one-tenth of a second, a listing of the occultation events to be observed with approximate timing predictions, a notepad and a red flashlight. The stopwatch should be set before each observing session, using either the telephone time signal or one of the reliable radio time signals that are broadcast. Bear in mind that some electronic stopwatches (even reputable makes) are liable to slow down or go completely awry in cold conditions.

The novice occultation observer is advised to start with disappearance phenomena behind the unlit leading western limb of the Moon. A degree of anticipation is in the observer's favour, since the star can be seen gradually approaching the edge of the Moon, especially when there is a hint of earthshine giving away the limb's position. Reappearances, especially at a brightly illuminated limb, require much skill and practice to record with anything like the level of accuracy more easily achievable with disappearance events.

People's reaction times vary. Everyone has a so-called personal equation which must be taken into account when making occultation timings. It has been shown that under good conditions disappearance events are recorded with an average delay of 0.3s. The effects of tiredness, coldness and discomfort will introduce further errors.

Lists of stars to be occulted in the near future are published in various astronomical ephemerides. Occultations of brighter stars, such as the Pleiades, Regulus and Spica, are

▲ *A drawing by Peter Grego of the waning crescent Moon amid the Pleiades, observed on 6 August 1988 using a 60 mm achromatic refractor.*

publicized in astronomical journals and magazines. Information about the star's magnitude and the predicted position angle (on the lunar limb) of disappearance and reappearance is provided. Position angle is measured from the north, anticlockwise around the lunar limb (i.e. through west first). It is vital to know the position angle, because you will need to concentrate upon just a small arc of the limb if you are to register the star's emergence the instant it happens.

Serious occultation work is exacting. It is essential that the observer knows the latitude and longitude of the observing site and its height above sea level. Observers in the United Kingdom can determine their exact grid reference from 1:2500 or 1:1250 scale Ordnance Survey maps. Observers in other parts of the world can often find such information on the internet: for example, from the US Geological Survey website on the Geographic Names Information System (GNIS) at http://geonames.usgs.gov/gnishome.html, or Geoscience Australia's place name search facility at http://www.ga.gov.au/map/names/.

Planets, being occupants of the near-ecliptic, are occasionally occulted by the Moon. Unlike stars, which disappear and reappear almost instantly, a planet – which is not a point source – will do so gradually. The time a planet takes to be occulted depends on its angular diameter and the angle at which the lunar limb approaches it; the angle varies somewhat with the observer's location on the Earth. The shortest occultations occur when a planet encounters the preceding or following limb square-on. At the other extreme, a planet can appear to skim along the north or south limb of the Moon,

PLANETARY OCCULTATIONS BY THE MOON, 2010 TO 2020			
Date	Planet	Moon's phase	Visibility
2010 Dec 6	Mars	Waxing crescent	Eastern USA
2011 Jun 30	Venus	Waning crescent	Central Asia (daylight)
2012 Jul 15	Jupiter	Waning crescent	UK, Europe
2012 Aug 14	Venus	Waning crescent	Far East
2014 Feb 26	Venus	Waning crescent	Central Asia (daylight)
2014 May 14	Saturn	Waxing gibbous	Australia
2014 Aug 4	Saturn	First quarter	Australia
2014 Oct 25	Saturn	Waxing crescent	UK, Europe
2015 Oct 9	Venus	Waning crescent	Australia
2015 Dec 7	Venus	Waning crescent	USA (daylight)
2016 Apr 6	Venus	Waning crescent	UK, Europe (daylight)
2019 Feb 2	Saturn	Waning crescent	UK, Europe, Asia
2019 Apr 25	Saturn	Waning gibbous	Australia
2019 Aug 12	Saturn	Waxing gibbous	Australia
2020 Feb 18	Mars	Waning crescent	Western USA
2020 Jun 19	Venus	Waning crescent	UK, Europe (daylight)

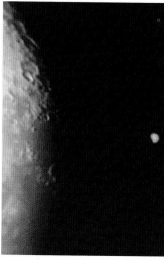

▲ Saturn emerges from occultation at the bright limb of the Moon on 16 April 2002. The rings are clearly visible.

▶ Jupiter is occulted at the unilluminated limb on 23 February 2002. Both photographs were taken by Paul Stephens.

producing a longer occultation, which may not include a period of totality. When partially hidden by a bright lunar limb, a planet often appears to be separated from the Moon by a dusky line – a simple contrast effect between the two bodies, and not an indication of a lunar atmosphere.

During the past few years there have been a number of spectacular planetary occultations. From the UK, for example, brilliant Venus underwent occultations in 2004, 2007 and 2008; Jupiter and its four bright satellites were occulted in 2002, while Saturn experienced occultations in 2001, 2002 and 2007. Having been fortunate to observe the Saturn occultations of 12 November 1997 and 16 April 2002, the disappearance of the ringed planet behind the Moon's dark limb and its reappearance at the bright limb were truly spectacular events. In the first event, Saturn took about a minute to disappear behind the lunar limb from one end of its ring system to the other; once the glare of Saturn is extinguished at the dark limb some of its fainter satellites and elusive F-ring may come into view.

Each year the vigilant amateur astronomer equipped with nothing grander than a 60 mm refractor will have the opportunity to make dozens of accurate occultation timings of stars brighter than seventh magnitude. Observers are encouraged to contribute their timings to any of the excellent organizations that compile occultation records, co-ordinate observations and distribute predictions. These include the Occultation Section of the Society for Popular Astronomy (SPA) and the British Astronomical Association (BAA), both in the UK, and the International Occultation Timing Association (IOTA).

THE SPACE-AGE MOON

Investigation of the Moon by spacecraft became a priority for both the United States and the Soviet Union in the late 1950s. In the final four decades of the 20th century, robotic lunar probes and the Apollo missions taught us more about the Moon than had been learnt during the previous four centuries of lunar observation.

Soviet Moon probes

The Soviet Moon programme was conducted mainly through a series of probes which bore the name Luna (though the first three were originally named Lunik). Two dozen of them were dispatched between January 1959 and August 1976, beginning with basic lunar flybys, and progressing through unmanned landers, roving vehicles and sample-return missions.

Luna 1 was launched in January 1959. The marksmanship on this historic first Moonshot was not bad – Luna 1 was off-target by just 5000 km. Nine months later, Luna 2 became the first probe to hit the Moon, blasting a 10-metre hole between craters Autolycus and Archimedes, followed shortly afterwards by the rocket's upper stage. To honour mankind's first true interplanetary contact, the area of impact was named Sinus Lunicus (Lunik Bay) by the International Astronomical Union in 1970. The area is easy to locate through small telescopes, although the craters marking the impacts are far too tiny to be seen.

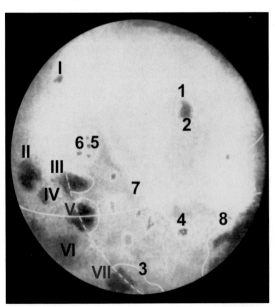

Nearside features:
I   Mare Humboldtianum
II Mare Crisium
III Mare Marginis
IV Mare Undarum
V Mare Smythii
VI Mare Fecunditatis
VII Mare Australe
Farside features:
1 Mare Moscovianum
2 Sinus Astronautarum
3 Mare Australe border
4 Tsiolkovsky
5 Lomonosov
6 Joliot-Curie
7 Soviet Mountains
   (non-existent)
8 Mare Somnii

◀ In 1959, the Soviet probe Luna 3 returned the first images of the farside of the Moon. This numbered map, released shortly after the mission, shows several nearside and farside features.

▶ *In February 1966 Luna 9 became the first probe to soft-land successfully on the Moon. It returned the first grainy images from the lunar surface. The rock-strewn lava plain of Oceanus Procellarum is clearly visible.*

A month later, Luna 3 obtained the first images of the Moon's farside, showing it to be dramatically different from the nearside. No large dark marial plains were visible, just a few crater-sized dark patches amid a brightly mottled expanse. Soviet astronomers analysed the fuzzy images, and a map was prepared which showed hundreds of farside features, with names for 18 of the most prominent formations. Noteworthy is Mare Moscoviense (Sea of Moscow) and the large, dark-floored plain of Tsiolkovsky (named after a Russian spaceflight theoretician) with its prominent central mountains. Because the first maps were based on poor images, many of the features they depict do not exist, most notably the 2000 km 'Soviet Mountain' range, later shown to be a vague alignment of patches.

After a series of failures, soft-landing success came in February 1966 when Luna 9 set down some 40 km north-east of the crater Cavalerius in Oceanus Procellarum. The landing area was named Planitia Descensus (Plain of Descent) to commemorate the event. Planitia Descensus is easy to locate through a telescope since it lies halfway between the brilliant Reiner Gamma and Cavalerius' northern rim. Luna 9's photographs were first made public by Bernard Lovell at the Jodrell Bank radio observatory in England, where the craft's transmissions had been intercepted. Luna 9 showed that the Moon's surface (in the neighbourhood of the craft, at least) was not, as some lunar scientists had theorized, covered with a thick layer of dust capable of completely engulfing heavier spacecraft – good news for the Apollo mission planners.

Luna 10 became the Moon's first artificial satellite in April 1966. During its two-month mission, the orbiting craft obtained valuable scientific data: the lunar magnetic field was found to be weak, and no radiation belts were discovered. Similar success was attained in lunar orbit by Lunas 11 and 12, both launched later that year.

In December 1966, Luna 13 soft-landed in Oceanus Procellarum, 40 km south-east of the crater Seleucus, and secured high-resolution television pictures and panoramic sweeps of the lunar landscape. The probe's extensible arms sampled the lunar soil (the regolith). The regolith was tested with an explosive hammer and found to be of similar density to terrestrial soil. The terrain in the craft's vicinity was flat and featureless, with only small craterlets visible.

In September 1970, Luna 16 soft-landed in north-eastern Mare Fecunditatis. A regolith sampling mechanism attached to a long extensible arm drilled holes into the surface; 100 grams of samples were deposited in the craft's upper stage. A day later the upper stage was launched Earthwards, leaving behind the descent stage to continue making measurements on the Moon's surface. This was the first successful robotic sample-return mission.

A large automated roving vehicle called Lunokhod was transported to the Moon in November 1970 on board Luna 17, which landed 60 km south-east of Promontorium Heraclides in Mare Imbrium. The eight-wheeled rover was controlled from Earth by a small team who navigated obstacles via a basic TV system. Four high-resolution TV cameras gave excellent views of the lunar topography. In its 322-day mission, Lunokhod traversed 10.5 km, going south across the plain and back to the landing area, taking 20,000 photographs and 200 panoramic pictures.

Luna 18 crashed in the uplands north-east of Mare Fecunditatis in September 1971. The next successful probe, Luna 20, landed five months later just a few kilometres from its predecessor's crash site, and returned to Earth 30 grams of soil. In January 1973, Luna 21 deposited Lunokhod 2 on the southern floor of the crater Le Monnier, adjoining Mare Serenitatis. Lunokhod 2 was a more advanced version of the first roving vehicle, with additional objectives including magnetic and photometric studies. In its lifetime of four months, Lunokhod 2 traversed 37 km to Le Monnier's southern ramparts and skirted the crater's edge, returning valuable lunar data. The last successful mission in the series was Luna 24, which returned a regolith sample taken from 2 metres below the surface of Mare Crisium, near the craterlet Picard X, in August 1976.

Lunar probes of the United States

America's early attempts to reconnoitre the Moon were beset by a string of disasters. From August 1958 to January 1964 no less than 15 spacecraft failed to meet their mission objectives in one way or another. The USA achieved success in robotic lunar exploration with three separate programmes – Ranger, Surveyor and Lunar Orbiter.

Ranger to the Moon The 2.5-metre-high Ranger craft were one-way missions intended to secure thousands of increasingly detailed images before their destructive impact on the lunar surface. For 17 minutes before impact on 31 July 1964, Ranger 7 returned pictures

▶ *On 24 March 1965, Ranger 9 imaged the crater Alphonsus (diameter 108 km) from a height of 442 km, about 3 minutes before it impacted on the crater's floor. More detail can be seen in this image than with the largest telescopes on Earth.*

of the area 100 km south-west of Fra Mauro – this area was later named Mare Cognitum (the Known Sea) to commemorate the achievement. As the probe neared the Moon, features beyond the resolution of the best telescopes on Earth came into view, and the final image, taken from a height of 1.5 km, showed features as small as a metre across. On 20 February 1965, Ranger 8 obtained high-resolution pictures of south-western Mare Tranquillitatis, finally impacting 60 km north-east of where Apollo 11 would land four and a half years later. The final Ranger returned images of the crater Alphonsus on 24 March 1965. Ranger 9's mission was a complete success – indeed, the American public shared the excitement as more than 200 pictures were broadcast live on national television. Ranger 9's final image was taken from 530 metres and showed features as small as 250 mm across. Alphonsus' rilles, which from Earth appear linear and uncomplicated, were seen to consist of chains of connected rimless circular depressions. This was the first indication that other seemingly simple linear rilles might not be as they appeared at low magnification through terrestrial telescopes.

Surveyor on the Moon From 1966 to 1968, the sophisticated Surveyor probes soft-landed at various locations on the Moon's near-side. Five out of the seven Surveyors were successful. The first of them, Surveyor 1, soft-landed on the flat plains of Oceanus Procellarum, some 40 km north of the crater Flamsteed, in June 1966. Through a telescope, the landing site appears to be in a relatively bland portion of the Moon – a plain broken here and there by the odd craterlet, small hill and low wrinkle ridge.

The next successful Surveyor, the third in the series, soft-landed in Mare Insularum in April 1967, 40 km south-east of where Luna 5 crashed. Through a telescope, the immediate vicinity of the landing site is fairly bland, though the spectacular craters farther afield – Copernicus to the north-west and Ptolemaeus to the south-east – make it easy to pinpoint. Surveyor 3 found the upper soil to have the properties of fine, damp sand, capable of supporting the weight of astronauts and their equipment. Two and a half years later, in a remarkable pinpoint landing, Apollo 12 touched down just 200 metres north-west of Surveyor 3.

In July 1967 Surveyor 4 failed just minutes from its planned landing. Two months later, and after several potentially disastrous malfunctions, Surveyor 5 landed 70 km south of Lamont, in Mare Tranquillitatis. As well as taking thousands of TV pictures, the probe analysed the regolith using an alpha-ray scatterer, a sensitive instrument capable of identifying the surface composition. The observations indicated that the regolith was a silicate material of basaltic composition. Surveyor 6 landed in Sinus Medii, a few kilometres north of the crashed Surveyor 4, in November 1967. The final Surveyor made its descent to a mountainous region only 20 km north of the rim of the prominent crater Tycho in January 1968 – the grandest and most testing touch-down point for any of the Surveyors, in the midst of the highly cratered southern uplands.

Orbiter round the Moon Between them, Lunar Orbiters 1 to 5 mapped 99% of the Moon to unprecedented accuracy, helping scientists to choose the landing sites for the later Surveyors and Apollo. In 77 days (527 orbits of the Moon) commencing in August 1966,

◀ 'The picture of the century' – an oblique view of the 100 km diameter crater Copernicus, taken by Lunar Orbiter 2 in November 1966. Telescopic observers see Copernicus directly from above, so this image is interesting to compare and gauge the real heights of the mountains and depth of the crater floor.

► In May 1967 Lunar Orbiter 4 imaged the impressive multiringed basin of Mare Orientale on the Moon's western limb. Telescopic observers can view the outermost mountain rings, and at a favourable libration can actually see the dark marial patch of Orientale.

Lunar Orbiter 1 photographed 5.2 million sq km of the lunar surface, including many prospective Apollo landing areas. The best known of the probe's pictures was the first image of the crescent Earth about to set behind the lunar limb. At the end of operations the probe was deliberately crashed on to the Moon's surface, as were subsequent Lunar Orbiters, to prevent it interfering with later Apollo missions.

Lunar Orbiter 2 began circling the Moon in November 1966, and took nearly 200 splendid pictures, the most famous of which was taken from a height of 20 km, some 140 km south of the rim of Copernicus. The view of the crater's blocky fringe and its jumbled central massif was acclaimed as 'the picture of the century'. Lunar Orbiter 2 also took the first *in situ* photographs of a lunar eclipse, on 24 April 1967.

The third Lunar Orbiter entered orbit in February 1967. So good was its photographic resolution that it picked out the diminutive Surveyor 1 probe resting on Oceanus Procellarum. The excellent images also included close-up views of Rima Hyginus in Sinus Medii – a firm favourite of telescopic rille observers – and the farside crater Tsiolkovsky.

Lunar Orbiter 4 became the first spacecraft to enter polar orbit around the Moon, in May 1967, its arrival making a record total of three operational US craft in lunar orbit. As the prospective Apollo sites were already well charted, the probe concentrated on the areas of the Moon that were inadequately mapped. A surprise discovery was made just past the Moon's south-eastern limb (out of the range of telescopic observers, sadly), where a huge 310 km long gash was seen cutting through the crater Sikorsky, to the north of the large formation Schrödinger. Stunning shots of the vast multiringed asteroidal impact basin Mare Orientale were obtained. Telescopic observers can view the basin's outermost mountain rings on the south-western lunar limb, and at favourable librations the dark marial patch of Mare Orientale is visible.

The final probe in the series, Lunar Orbiter 5, entered a near-polar orbit in August 1967 and returned hundreds of high-quality images. The probe's speed variations in lunar orbit were analysed and found to be caused by the gravitational effects of concentrations of mass (known as mascons) beneath the lunar maria. Excellent views were also obtained of the lunar eclipse of 18 October.

By rocket to the Moon

Between July 1969 to December 1972, twelve US Apollo astronauts walked on the Moon. The Apollo programme helped to build a picture of the various internal and external processes that sculpted the lunar surface.

Apollo 11 – historic footsteps The first landing was made on level terrain near the lunar equator, ensuring the best possible arrangement for radio communication, landing and lift-off. The site chosen was a small area in southern Mare Tranquillitatis – just beneath the left eye of the 'man in the Moon'. The site, near Tranquillitatis' southern shore, is of considerable interest. To the north, wrinkle ridges radiate from a ghost crater called Lamont, a feature visible only when illuminated by a low Sun. Various small craters, domes and rilles in the southern part of Mare Tranquillitatis attest to its interesting geological history.

At 21h 18m Universal Time on 20 July 1969, Neil Armstrong calmly announced 'Houston, Tranquillity base here. The *Eagle* has landed.' Six hours after landing, Armstrong swung open the lunar module (LM) hatch and descended by a ladder attached to one of the craft's legs. The descent was made difficult because the Sun was directly behind the LM, casting a pitch-black shadow over the surroundings. Armstrong's famous first words as he stepped on to the grey lunar soil were appropriate and poetic: 'That's one small step for a man – one giant leap for mankind.' He was joined on the surface by Buzz Aldrin, who described the Moon as a scene of 'magnificent desolation'. Humanity's historic first walk on another world, over a few hundred square metres of the lunar surface, lasted nearly two and a half hours. Several scientific experiments were set up, including a seismometer to measure Moonquakes and a laser reflector for measuring the Earth–Moon distance. One of the astronauts' main priorities was to collect samples of lunar rock and soil for geologists to analyse. In all, 21 kg of material was harvested from a small area due west of the LM.

Apollo 12 – encounter with Surveyor On 18 November 1969, Charles Conrad and Alan Bean landed Apollo 12's LM, *Intrepid*, on Oceanus Procellarum some 100 km south-east of the crater Lansberg, at a site just 200 metres from the Surveyor 3 probe that had arrived over two years before. Through a telescope, the Apollo 12 site is difficult to pinpoint with certainty, as there are no prominent features nearby. To the north-west lie the remnants of a large crater buried by lava flows and identifiable only by the presence of a few low hills. There is little telescopic evidence of faulting in the immediate area. The region was probably covered to a considerable depth by material ejected from the Copernicus impact site around 900 million years ago.

Conrad and Bean made two Moonwalks. On the first they ventured a short distance west of *Intrepid*, passing two strange-looking mounds of lunar soil on the way. The Apollo Lunar Scientific Experiment Package (ALSEP) was assembled, and rock and soil samples were taken. During their second excursion the astro-

▼ *Buzz Aldrin took this image of a footprint in the soil at Tranquillity base – a simple shot, but one of the most famous images returned by Apollo.*

▼ *This image of the Apollo 11 landing site in south-western Mare Tranquillitatis was taken from the lunar module before its descent to the Moon's surface on 20 July 1969. Tranquillity base is near the morning shadow line, just to the right of centre. At lower right is the 23 km diameter crater Maskelyne. The large black object at lower left is not a shadow but a lunar module thruster in the camera field of view.*

◄ *Apollo 12 astronaut Alan Bean examines Surveyor 3. The astronauts walked to the spacecraft from their own lunar module (visible on the horizon), which landed just 200 metres away.*

nauts rounded the rim of the 150-metre Head Crater, continued another 300 metres south to the 50-metre Bench Crater, examining an exposed rock outcrop and 'melted' central elevation, then proceeded to the diminutive Sharp Crater with its rays. They returned to *Intrepid* via Surveyor 3, which by now was coated with a thin layer of dust. A quantity of the craft's tubing was snipped off, and its soil scoop and TV camera were salvaged to see how the materials had fared in the harsh lunar environment.

Once the crew had safely departed from the Moon and transferred into the command module, the unwanted ascent stage was allowed to fall to the surface to determine the seismic effects of a known impact event. The 2-tonne craft smashed into the lunar surface 60 km from the ALSEP at a speed of 6000 km/h, carving out a sizeable crater. The seismometer recorded reverberations lasting for 55 minutes as the Moon 'rang like a bell'.

Apollo 14 – Fra Mauro finally In April 1970 the crater Fra Mauro had been the intended landing site of Apollo 13, but an explosion in the service module halfway to the Moon led to the mission being aborted. More than nine months later, Apollo 14's LM *Antares*, manned by Alan Shepard and Edgar Mitchell, touched down on an 8° slope among the hills near Fra Mauro on 31 January 1971, 180 km east of the Apollo 12 site. Through a telescope the Apollo 14 site is not difficult to locate, just 20 km north of the wall of Fra Mauro. A low Sun shows the area to be covered with a series of low ridges running north–south which may have been produced around 3.8 billion years ago by the Imbrium impact. The area is rich in tonal variations. Fra Mauro is an ancient crater some 100 km across, its floor crossed by clefts and ridges. The craters Bonpland and Parry are connected to the south wall of Fra Mauro, and all three craters have been flooded with lava.

Shepard and Mitchell tried to walk up the sloping rock-strewn flanks of the 400-metre Cone Crater, 1.5 km from *Antares*. They got to within 30 metres of the rim (though they did not know how close they had been) before time and tiredness (Shepard was 47) forced them to abandon the exercise. The discovery of white boulders near Cone Crater caused a flurry of excitement. These rocks were breccias – amalgams of large angular crystals cemented together by a fine crystalline matrix. Such material provides clues to the various impact events in the Moon's history. The cratered lunar highlands are composed of many different types of breccia. Those containing just one type of rock indicate (logically enough) an impact on a crust made up of a single rock type. Other breccias can contain mixtures of various rock types and a wide variety of fragment sizes, and their composition reflects the nature of the surface that was impacted and at what stage of the Moon's history the impact event that produced the breccia took place.

Apollo 15 – viewing a valley The lunar Apennine mountains are an easy target for the telescopic lunar observer. The peaks of this vast mountain range bordering Mare Imbrium cast pointed black shadows on to the grey lava plains around first and last quarter. Apollo 15 was the first mission to explore a boundary between lunar seas and mountains, in addition to a sinuous rille. This feature, Rima Hadley, is over a kilometre wide, tens of metres deep and around 80 km long, and it can be seen through a 150 mm reflector.

The Apollo 15 lunar module, *Falcon*, landed near Mons Hadley on 30 July 1971. Astronauts David Scott and James Irwin set out the experiments, which included the lunar seismic equipment – a pyrotechnic 'thumper' and mortar box. They enjoyed a significant advantage over the crews of previous missions – they had a Lunar Roving Vehicle (LRV), known as the Moon buggy. Their first drive took them due south over the flattish mare floor and along the rim of Rima Hadley to the base of Montes Apenninus. There they expected to find large chunks of excavated bedrock, but the sizeable boulder they encountered was thought to have come from the lava-flooded mare. Having put the LRV through its initial paces and collected lunar samples *en route*, the returning astronauts deployed the important ALSEP equipment.

The second LRV excursion took the astronauts south-east to the base of the mountains past a group of craters known as South Cluster. Here, the layering of rocks was examined to discover what types of lava had flowed in the past. Rocks with bubbles in them were discovered, some with holes several centimetres across and resembling Swiss cheese. Such odd-looking material is thought to have solidified from a particularly frothy kind of lava. Other volcanic rocks were glassy in lustre, suggesting that they cooled very rapidly immediately after erupting on to the surface. Various types of breccia were found, some pure white and others black and white.

During the eventful third trip in the Moon buggy, great interest was aroused by the discovery of green-tinted soil. Analysis back on Earth suggested that this vividly coloured material, composed of microscopic glassy spheres, was formed in a fountain-like eruption of lava. Rima Hadley was visited for the second time on this final Moon drive. Substantial piles of debris could be seen at the base of the rille, comprising soil and angular boulders up to 30 metres across. Bedrock was exposed along the walls of the rille, and the samples taken here were the only ones obtained of such ancient material in the entire Apollo programme. One rock sample, taken from the top of a mound of lunar soil, came to be known as the Genesis Rock. It is almost pure anorthosite, a type of rock that made up most of the Moon's ancient crust, and is thought to be around 4.5 billion years old.

Before leaving the Moon, Scott re-enacted Galileo's demonstration that bodies of different masses will fall to the ground with the same acceleration due to gravity. The objects Scott chose were his geological hammer and a falcon's feather. TV viewers saw both objects fall to the surface from the same height at identical speeds in the Moon's near-vacuum, low-gravity environment. The spectacular LM ascent stage blast-off was recorded by the TV cameras of the LRV, which tracked the craft as it rose into the black sky. On the way back to Earth, the crew observed the total lunar eclipse of 6 August, describing the deep orange and red colours visible in the eclipsed portion of the Moon.

◀ Apollo 11 image of the near-full Moon (from their vantage point) taken from a distance of 18,000 km during the journey back to Earth. The dark oval area above centre is Mare Crisium, and to its left is Mare Tranquillitatis. Terrestrial observers can never view such a scene, as a considerable portion of the far side is shown at right, including the brilliant rayed crater Bruno, near the right edge at about 2 o'clock.

Apollo 16 – visiting Descartes A typical highland region north of the crater Descartes was chosen for Apollo 16. On 16 April 1972, the LM *Orion* carrying John Young and Charles Duke landed on the wall of a large unnamed eroded crater 50 km north of Descartes. This region is quite difficult to observe telescopically as it is full of disintegrated crater remnants. It is best to identify the prominent craters Albategnius and Theophilus and then locate Descartes in between. Some scientists thought that the region had been a site of extensive crater-making volcanism, but this was disproved when the returned samples were analysed on Earth. Lunar ray material collected here was found to be just 2 million years old, extremely young by lunar standards. This material, ejected by the South Ray crater impact, was some of the youngest discovered during the Apollo landings. There is a dearth of volcanic material near Descartes, and this crater, along with all the others in the area, is an ancient impact feature.

Orion touched down between South Ray and North Ray crater, 270 metres north of the target spot. On their second excursion, Young and Duke raced up Stone Mountain in the LRV and were treated to a wonderful panorama, in which light crater rays stretched across the lunar surface. On their third Moon drive they examined North Ray crater, 1 km across. Its interior was seen to be layered, with two distinct rock types visible. A giant rock on the crater's outer flank appeared to have been excavated by the impact that formed North Ray, about 50 million years ago.

Apollo 17 – lunar farewell Apollo 17, the last mission to carry humans to another world, visited an area 30 km south of the crater Littrow, near the south-eastern shore of Mare Serenitatis. This is another area bordering a highland and a mare, and photographs taken from orbit had revealed numerous craters encircled by mysterious dark haloes. About 30 km west of the landing site lie the rilles of Rimae Littrow, which intersect the concentric rilles around the border of Mare Serenitatis near the crater Clerke. These rilles can be observed through a small telescope under favourable conditions. The mountainous area appears markedly striated.

On 11 December 1972 Apollo 17's LM, *Challenger*, descended into the Taurus-Littrow valley. The last Moonwalkers were Eugene Cernan and Harrison Schmitt, Schmitt being the only qualified geologist to have visited the Moon. They drove their LRV towards a large blocky mountain called South Massif, across light-coloured material deposited by a lunar avalanche perhaps triggered by the Tycho impact 100 million years ago. Bedrock was exposed on the mountainside where it had shaken off its topsoil. Boulders had rolled down the slopes, leaving trails. Rocks brought back from this location were similar to those from the Apollo 16 site, but some specimens were found to contain material dating back almost to the Moon's birth more than 4.5 billion years ago, having been deposited on to the mountain by a deep impact much farther afield. A surprise discovery was orange soil. Initially thought to be a sign of recent volcanism, it turned out to consist of microscopic glassy beads which had probably been produced by a meteoritic impact. Towards the end of the final excursion, Cernan and Schmitt visited the Sculptured Hills – features believed to be piles of ejecta deposited around 4.2 billion years ago by the Serenitatis basin impact.

It was hoped that at least two more Apollo lunar landings would take place after Apollo 17. In 1970, NASA Administrator Thomas Paine announced budget cuts, and later cancelled the Apollo 18 and 19 missions. Apollo had fulfilled its political objectives the very moment Armstrong's boot imprinted its pattern in the lunar dirt. We can still visit the six Apollo landing sites with the aid of our telescopes, viewing them as though we were peering through the porthole of our very own spacecraft in lunar orbit.

Return to the Moon

Clementine – triumph of technology and thrift   The Clementine probe, a joint US Department of Defense/NASA mission, marked a resumption of lunar exploration after nearly two decades. Between February and May 1994, it began a comprehensive survey of the Moon from an eccentric polar orbit. Clementine obtained an incredible amount of new data about the Moon's surface – its topography, structure, composition and mineralogy – and more than 2 million lunar images were obtained by the probe. Clementine's four cameras obtained images at visual, infrared and ultraviolet wavelengths – a powerful combination which revealed previously unseen structure and composition. During each five-hour lunar orbit, up to 5000 images were transmitted.

Clementine's ranging system bounced laser signals off the Moon to accurately gauge its height above the lunar surface. At the lowest point in its eccentric orbit, the distance to the surface could be determined to within 50 metres. By combining this information with Clementine's detailed lunar images, three-dimensional models of the topography were computed. The Moon was found to have a topographic range of some 16 km, from the floors of the deepest craters to the tips of the highest peaks, with distinct differences between the nearside and the farside. Clementine's topographic map shows that a large area of the farside is occupied by a vast raised plateau, on average some 4 km higher than the nearside highlands.

All the nearside maria were found to be depressed beneath the mean surface level. Mare Crisium was found to be the deepest large mare, some 4 km beneath the mean surface level. Mare Humboldtianum, near the north-eastern lunar limb, is revealed as an extensive basin whose central lava plain sits some 6 km beneath the mean lunar surface level. On the eastern limb of the Moon, Mare Marginis is a patchwork of lava flows averaging around 3 km deep, while immediately to the south, Mare Smythii is a more discrete circular basin a little deeper than its neighbour. The Schiller Annular Plain, an ancient impact feature 350 km in diameter near the south-west limb, does not show up on the Clementine topographic maps as clearly as might be expected – at the right illumination, this feature is an impressive sight to the terrestrial observer. The elongated crater Schiller on the plain's north-eastern border is far deeper.

At first sight, the farside basins – with the exception of the multiringed bull's-eye of Mare Orientale – do not appear as large or as dramatic as those of the nearside. But Clementine discovered the true extent of the farside basins. When the farside southern region, occupied by the dark-floored craters Leibnitz, Von Kármán, Poincaré, Apollo and Mare Ingenii, was charted, it was found to be wholly contained within a vast circular depression more than 2000 km across and averaging 5 km deep. Named the South Pole–Aitken basin, this feature is the Solar System's largest known impact scar. It had been seen before, most recently in 1990 by the Galileo spaceprobe on its way to Jupiter, but Clementine gave us hard topographical

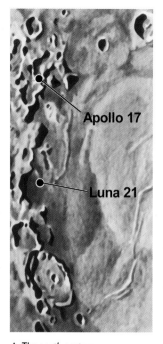

▲ *The south-eastern shore of Mare Serenitatis was the landing site of the last manned lunar mission, Apollo 17, in December 1972, and two months later the Soviet probe Luna 21. This view is based on an observation by Peter Grego.*

information and detailed images which revealed its sheer size. It is also the deepest impact basin in the Solar System, in places going down to 12 km beneath the mean farside surface level. It is no exaggeration to claim that the asteroidal collision that formed the South Pole–Aitken basin more than 4 billion years ago almost destroyed the Moon. A vast quantity of material from the Moon's mantle was churned up and erupted on to the surface as a result of the collision, and Clementine showed that the basin has a unique surface composition, rich in iron and titanium.

It has long been suspected that certain areas at the Moon's north and south polar areas lie in perpetual shadow. These are the floors of craters deep enough not to have been illuminated by the Sun's rays for many millions of years. But from the Earth, the Moon's near-polar regions are difficult to map, especially the crater-packed southern uplands. Apollo's cameras barely extended our knowledge of the north and south poles, since they made their lunar surveys from orbits over the Moon's equator, and only viewed the polar regions at an oblique angle and from some distance when they were *en route* to the Moon or returning to the Earth. Before Clementine, an area of 'luna incognita' in excess of 100,000 sq km existed at the south pole, extending from the craters Amundsen to de Roy. Clementine mapped the entire Moon from polar orbit and found that the Moon's south pole was indeed pitted with features deep enough to experience eternal lunar night. The largest such shadow zone lies over the south pole immediately to the west of the crater Amundsen and measures over 100 km in diameter.

Radar signals beamed at the south polar shadow zones were reflected far more strongly than those from the rest of the Moon's surface. Scientists stressed that the radar reflections did not represent 'a clearly unambiguous icy signature', conscious that the reflections may have been produced by a very smooth area of dense rock. However, large-scale areas of smooth, bare rock do not exist elsewhere on the Moon. Scientists remained cautious, although many were sure that the enhanced radar echoes represented the tell-tale signs of large patches of water ice lying in the south polar shadow zone on crater floors, permanently at −180°C.

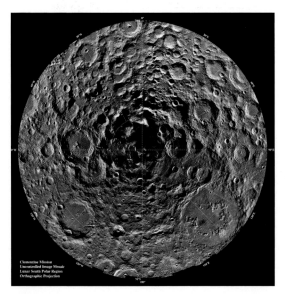

◄ A mosaic of 1500 Clementine images of the Moon's south polar region. The south pole is at centre, and the image extends to 70°S at the edge (image diameter about 1250 km). The permanently shadowed craters at the south pole are thought to harbour vast quantities of water ice. The top half of the image shows the lunar nearside, but at favourable librations, observers can glimpse many of the features below the centre line.

Clementine Mission
Uncontrolled Image Mosaic
Lunar South Polar Region
Orthographic Projection

Lunar Prospector Launched in January 1998, NASA's final Moon probe of the 20th century made the search for lunar water an official quest. Onboard, a gamma-ray spectrometer measured the composition of the Moon's surface and mapped the distribution of a number of chemical elements. Lunar Prospector's neutron spectrometer specifically detected hydrogen, whose presence in large quantities seemed to indicate the existence of water ice within deep shadow-filled craters at the Moon's south pole. In July 1999 at the end of its operational lifetime, the probe was deliberately crashed near the lunar south pole; it had been hoped to detect traces of water in the plume thrown up by the impact, but no trace of ejected debris was observed.

If lunar ice does exist, it probably arrived over the aeons in sporadic cometary packages. Each impact will have vaporized the comet's nucleus – a city-sized 'dirty snowball' – producing large clouds of dust and gas which temporarily swathed the Moon. The gases would have frozen and been precipitated in a snowfall over the Moon's cold dark side, but this material would not have survived on the Moon's surface once sunlight had warmed it. Most of the snow would have sublimated (changed from a solid state to gas), and escaped into space. But a tiny proportion of the comet's ices would have survived if it had been deposited in the cold polar shadow zones. Layers of comet snow mixed with lunar rock and soil will have accumulated over millions of years.

In June 2009 NASA dispatched Lunar Reconnaissance Orbiter (LRO), the United States' first Moon probe in more than a decade. From its 50 km orbit, LRO is surveying resources and identifying candidate landing sites for human missions, as well as providing incredibly detailed high-resolution images; these include images of the landing stages and some of the trails and equipment left on the Moon by the Apollo programme. LCROSS (a satellite launched from LRO) intends to undergo a controlled impact in a shadowed crater at the Moon's south pole in order that the plume of ejected material can be analysed; this may reveal the presence of water (hinted at by Lunar Prospector) which would be an incredibly useful resource.

International interest in probing the Moon All the world's space agencies are now taking advantage of new technology in spacecraft design and instrumentation to probe the Moon. The European Space Agency's SMART-1 began its solar-powered ion thruster glide from Earth orbit in September 2003, eventually reaching lunar orbit in November 2004. Once there, SMART-1 used infrared imaging to search for lunar water ice, while the entire Moon's chemical composition was charted using an X-ray spectrometer. Like Lunar Prospector, the probe was crashed into the Moon at the end of its mission, and an impact flash was observed and imaged. Observations did not prove the existence of water ice on the Moon, and there remains an active debate about the origin and form taken by the hydrogen detected at the lunar poles.

Japan's SELENE (Selenological and Engineering Explorer) lunar orbiter (nicknamed Kaguya) was launched in September 2007. Throughout its two-year period of operations the probe imaged the surface with a high-resolution terrain camera and high-definition video camera, probed the surface with a radar sounder and laser altimeter, and determined surface composition using an X-ray fluorescence spectrometer and a gamma-ray spectrometer. SELENE also released two subsatellites, one to relay radio signals and the other to measure the Moon's gravity field.

Burgeoning spacefaring superpower China entered its first phase of lunar exploration in November 2007 with the orbital probe Chang'e 1. Among its goals were to produce detailed 3D images of the lunar surface, to determine lunar geology and map the distribution of various chemical elements.

India's Chandrayaan-1 entered lunar orbit in November 2008, where it conducted scientific investigations. Contact with the probe was lost in August 2009, but the following month it was announced that data gathered by the probe's Moon Mineralogy Mapper (a NASA instrument) revealed extensive amounts of water in the lunar soil, proving that large areas of the Moon's surface are 'damp' – probably the result of oxygen in the soil combining with hydrogen in the solar wind. You could 'squeeze' a litre of water from each cubic metre of lunar topsoil! India plans a follow-up mission in 2012, Chandrayaan-2, which will entail landing a motorized Moon rover.

Future prospects A host of lunar probes are scheduled to be dispatched within the next few years: GRAIL (a gravity mapping satellite), LADEE (lunar atmosphere and dust experimental satellite), plus ILN Node 1 and ILN Node 2 (surface scientific stations). Russia's Luna-Glob 1 (a seismic impactor) and Luna-Glob 2 (a south polar roving vehicle), plus follow-ups by China (Chang'e 2) and India (Chandrayaan-2) are all part of the Lunar Precursor Robotic Program, providing detailed studies and resources prior to a human return to the Moon.

Human return NASA's Constellation programme is set to take over from the Space Shuttle to carry forward the United States' human presence in space. The programme will see the development of a number of spacecraft and boosters that will send astronauts to the Moon by 2018, and onwards to Mars in the ensuing years. Two main boosters, the Ares I and the bigger Ares V, plus the Earth Departure Stage (EDS), will be the rocket workhorses of the early Constellation programme. Crew vehicles under development are the Orion capsule and the Altair lunar lander – essentially scaled-up versions of the Apollo Command and Lunar Modules.

GLOSSARY

albedo A measure of an object's reflectivity. A pure white surface reflecting 100% of light falling on it has an albedo of 1.0; a pitch-black, totally unreflective surface has an albedo of 0.0.

anomalistic month The time taken for the Moon to complete an orbit from one perigee to the next – 27 days, 13 hours and 18.5 minutes.

aperture The diameter of a telescope's objective lens (for a refractor) or primary mirror (for a reflector).

apogee The point in the Moon's orbit where it is farthest from the Earth. The Moon's maximum distance from the Earth at apogee is 406,700 km.

basalt A dark, fine-grained volcanic rock, low in silicon, with a low viscosity when molten. Basaltic material – lava which welled up from the interior after impacts by asteroids, and later solidified – fills many of the Moon's major basins, especially on the nearside.

basin A very large circular impact feature (often with a structure of multiple concentric rings), usually lava-flooded. The largest and most conspicuous basins on the Moon are on the nearside, and most are filled to their outer edges with basalts. The farside basins are generally smaller and are lava-flooded mainly at their centres only.

caldera A sizeable depression in the summit of a volcano, caused by subsidence or explosion.

central peak An elevation at the centre of an impact crater, usually formed by elastic rebound of the lunar crust after impact.

cleft A small rille.

crater A circular feature consisting of a largely level floor, often lower than the surrounding terrain, bounded by a circular (or near-circular) wall. Almost all the large craters visible on the Moon were formed by asteroidal impact, but a few smaller craters are *endogenic*, of volcanic origin.

craterlet A tiny crater on the verge of telescopic resolution.

crescent Moon The period between new Moon and dichotomy when the Earth-turned lunar hemisphere is less than half-illuminated.

dark side (night side) The hemisphere of the Moon not experiencing direct sunlight.

dichotomy Half-phase (first quarter or last quarter Moon).

dome A low, rounded elevation with shallow-angled sides. Most domes are volcanic in origin, but others may have been formed by subcrustal pressure.

earthshine The faint blue-tinted glow of the Moon's unilluminated hemisphere, visible with the naked eye when the Moon is a narrow crescent. It is caused by sunlight reflected on to the Moon by the Earth.

ecliptic The apparent path of the Sun on the celestial sphere. The ecliptic is inclined by 23.5° to the celestial equator. The major planets follow paths close to the ecliptic, and the Moon's path is inclined by some 5° to it.

ejecta Material that is thrown out from the site of a crater-forming impact and lands on the surrounding terrain. Large impacts produce ejecta sheets composed of melted rock and larger solid fragments, in some cases ejecting streams of material now visible as bright *ray* systems.

ellipse The shape of closed celestial orbits: a closed curve with two focal points on its main axis. The Earth lies at one focal point of the Moon's orbital ellipse.

elongation The angular distance of the Moon or a planet from the Sun, viewed from the Earth, measured between 0° to 180° east or west of the Sun. For example, the first quarter Moon has an eastern elongation of 90°.

endogenic Having an internal origin. Lunar volcanoes and faults are endogenic.

ephemeris A table of calculated times, positions and other quantities relating to some future astronomical event such as the rising and setting times of the Moon.

farside The hemisphere of the Moon that is turned away from the Earth. The farside relates to all features between 90° east and 90° west, but libration allows the terrestrial observer to glimpse some 59% of the entire lunar surface over time.

fault A crack in the crust caused by tension, compression or sideways movement.

first quarter The half-phase between new Moon and full Moon, occurring one-quarter of the way into the lunation.

full Moon When the lunar disk is completely illuminated by the Sun. Viewed from above, the Sun, Earth and Moon are in line.

gibbous The phase of the Moon between dichotomy and full.

graben (plural graben) Geological term for a valley bounded by two parallel faults, caused by crustal tension. Most linear *rilles* are graben.

highlands (uplands) Heavily cratered regions of the Moon of generally higher elevation than the maria. They appear significantly brighter than the maria.

impact crater A hole in the Moon's crust formed by a large projectile striking the Moon at high speed, causing either a mechanically excavated pit (meteoroid impacts) or a large explosive excavation (asteroid impacts).

last quarter The half-phase between full Moon and new Moon, occurring three-quarters of the way into the lunation.

lava Molten rock extruded on to the surface by volcanic activity.

libration The apparent rocking motion of the Moon around its axis that allows observers to glimpse a total of 59% of the lunar surface over a period of time.

limb The edge of the Moon's disk.

lithosphere The Moon's solid crust.

lunar eclipse Darkening of the Moon as it moves through the Earth's shadow. Lunar eclipses can be *penumbral*, *partial* or *total*, and happen at full Moon if the Sun, Earth and Moon are almost exactly in line.

lunar geology The study of the lunar rocks and the processes that sculpted the Moon's surface; an older term is *selenology*.

lunation A complete cycle of lunar phases, from one new Moon to the next, averaging 29 days, 12 hours and 44 minutes. This is the Moon's synodic month.

mare (plural maria) A large, dark lunar plain. Maria fill many of the Moon's large multiringed basins and comprise a total of 17% of the Moon's entire surface area.

massif A large mountainous elevation, usually a group of mountains.

nearside The hemisphere of the Moon that is constantly turned towards the Earth.

new Moon The lunar phase during which all of the nearside is unilluminated.

occultation The disappearance or reappearance of a star or planet behind the lunar limb.

perigee The point in the Moon's orbit where it is closest to Earth. The Moon's minimum distance from the Earth at perigee is 356,400 km.

ray A bright linear streamer radiating from a lunar crater, part of a crater's *ejecta* system.

regolith The lunar soil – a mixture of fine dust and rocky debris, produced by aeons of relentless meteoritic erosion.

rift valley A *graben*-type feature caused by crustal tension, faulting and vertical slippage of the middle crustal block.

rille A narrow valley. *Linear rilles* are caused by crustal tension and faulting. *Sinuous rilles* are the result of rapid erosion by fast-moving lava flows.

scarp A cliff produced by crustal tension, faulting and relative vertical movement between the two crustal blocks.

secondary cratering Craters produced by the impact of large pieces of ejecta from a large impact. Secondary craters often occur in distinct *chains*, where masses of ejected material following similar trajectories impacted almost simultaneously.

seeing A measure of the quality and steadiness of the atmosphere as it affects an image seen through the telescope eyepiece. Bad seeing is caused by atmospheric turbulence, largely through thermal effects.

selenology An older term for *lunar geology*.

synodic month The period taken for the Moon to complete one cycle of phases, from one new Moon to the next. It averages 29 days, 12 hours and 44 minutes

terminator The line separating the illuminated and unilluminated hemispheres of the Moon. From new Moon to full we observe the morning terminator. From full to new Moon we see the evening terminator.

volcano An elevated feature built up over time by the eruption of molten lava and ash. Lunar volcanoes are usually low, with shallow slopes and topped by tiny summit craters (vents). Volcanic activity on the Moon ceased more than 2 billion years ago.

walled plain A crater larger than about 70 km across with low walls and a flattish, often flooded floor.

waning Moon The period from full Moon to new Moon when the illumination of the visible lunar surface decreases.

wrinkle ridge A linear or sinuous feature of low elevation. Wrinkle ridges traverse many of the marial plains. Some are lava flow fronts, others are features formed by compression as the mare surface contracted, while a few trace the buried outlines of features such as craters or inner basin rings.

waxing Moon The period from new Moon to full Moon when the illumination of the visible lunar surface increases.

RESOURCES

Societies, groups and useful Internet resources

The Society for Popular Astronomy (SPA)
Website http://www.popastro.com (online joining facility)
A UK-based astronomical society founded in 1953, aimed at amateur astronomers of all abilities. Its publications include the quarterly magazine *Popular Astronomy*. The SPA has various observing sections, including an active Lunar Section (directed by the author since 1984) with its own journal, *Luna*.

The British Astronomical Association (BAA)
Website http://www.britastro.org/main/index.html
A UK-based astronomical association catering for more advanced amateurs. The BAA Lunar Section has an active Lunar Topographic Subsection directed by Colin Ebdon, and its own journal, *The New Moon*.

Association of Lunar and Planetary Observers (ALPO)
Website http://www.lpl.arizona.edu/alpo/
Founded in 1947, ALPO is dedicated to encouraging research and observations of all Solar System objects, both professional and amateur. It has an active Lunar Section.
ALPO Lunar Section Website
http://www.lpl.arizona.edu/~rhill/alpo/lunar.html

Chuck Taylor's Lunar Observing Group on Yahoo
Website http://groups.yahoo.com/group/lunar-observing/
An active web-based community for discussing all things lunar.

Consolidated Lunar Atlas
Website http://www.lpi.usra.edu/research/cla/index.shtml
Superb pre-Apollo photographic coverage of the Moon through large Earth-based telescopes.

Digital Lunar Orbiter Photographic Atlas of the Moon
Website http://www.lpi.usra.edu/research/lunar_orbiter/
A superb searchable web archive of the best images returned by the US Lunar Orbiter probes of the late 1960s.

Geologic Lunar Research Group
Website http://digilander.libero.it/gibbidomine/
A resource for lunar observers interested in lunar geology and research into lunar transient phenomena (LTPs).

Lunar and Planetary Institute (LPI)
Website http://www.lpi.usra.edu/lpi.html
The LPI, based in Houston, Texas, is a focus for academic participation in studies of the Solar System. It has extensive collections of lunar and planetary data.

NASA Lunar Exploration
Website http://nssdc.gsfc.nasa.gov/planetary/lunar/apollo_25th.html
Links to descriptions of all the NASA lunar missions.

The American Lunar Society
Website http://otterdad.dynip.com/als/
A group dedicated to amateur lunar studies, with numerous observing projects.

The Lunar Observer
Website http://users.adelphia.net/~dembowski/
An independent newsletter for students of the Moon. It is edited by Bill Dembowski, who is President of the American Lunar Society.

Unione Astrofili Italiani (UAI) Lunar Section
Website http://www.uai.it/sez_lun/english.htm
An active observing section, with an English version of its website.

Selected lunar bibliography

Some classic lunar works

Fauth, Philip, *The Moon in Modern Astronomy*, A. Owen & Co., 1909. 160pp. Beautifully illustrated, but contains some peculiar speculation about lunar ice.

Fielder, Gilbert, *Structure of the Moon's Surface*, Pergamon Press, 1961. 266pp. Detailed scientific speculation on the origin of lunar features in the pre-Apollo era.

Fielder, Gilbert, *Lunar Geology*, Lutterworth Press, 1965. 184pp. Detailed scientific speculation on the origin of lunar features in the pre-Apollo era.

Firsoff, V. A., *Strange World of the Moon*, Hutchinson, 1959. 226pp. Contains much interesting pre-Apollo speculation on the nature of the Moon.

Firsoff, V. A., *Surface of the Moon*, Hutchinson, 1961. 128pp. More lunar facts, speculation and theory.

Firsoff, V. A., *The Old Moon and the New*, Sidgwick & Jackson, 1969. 264pp. The last major lunar book written before the Apollo 11 landing.

Harley, Timothy, *Lunar Science*, Swan Sonnenschein, Lowrey & Co., 1886. 89pp. A lovely little book that waxes lyrical about the Moon's wonders.

Nasmyth, James, and Carpenter, James, *The Moon*, John Murray, 1885. 214pp. A Victorian classic with wonderful illustrations.

Proctor, Richard A., *The Moon*, Longman, Green & Co., 1886. 314pp. A very well-written guide to the Moon by one of the greatest Victorian popularizers of astronomy.

Spurr, Josiah, *Geology Applied to Selenology*, The Science Press, 1944. 112pp. An in-depth look at the Imbrian plain, with emphasis on lunar volcanism.

Wilkins, H. P., *Our Moon*, Muller, 1954. 180pp. A nice little guide to the Moon, full of lively descriptions.

Popular lunar books

Alter, Dinsmore, *Pictorial Guide to the Moon*, Thomas Y. Crowell Company, 1963. 183pp. A fine selection of telescopic photographs of the Moon, with lively descriptions.

Cherrington, Ernest H. Jr, *Exploring the Moon Through Binoculars and Small Telescopes*, Dover, 1984. 229pp. A well-written guide to the Moon and the features visible throughout a lunation.

Hill, Harold, *A Portfolio of Lunar Drawings*, Cambridge University Press, 1991. 240pp. Superb observations of a variety of lunar features by one of the world's most accomplished amateur lunar observers. Recommended.

Kopal, Zdenek, *The Moon*, Chapman & Hall, 1960. 131pp. A well-written general guide to the Moon from an early space age perspective.

Light, Michael, and Chaikin, Andrew (editors), *Full Moon*, Jonathan Cape, 1999. 243pp. Artist and photographer Michael Light drew on NASA's archive to put together an archetypal lunar journey in images, from take-off to landing. Many fabulous images of the lunar surface.

Moore, Patrick, *Guide to the Moon*, Lutterworth Press, 1976. 320pp. A well-written, uncomplicated guide to all aspects of the Moon.

Price, Fred W., *The Moon Observer's Handbook*, Cambridge University Press, 1988. 309pp. Some sound historical background on lunar observing, and much helpful advice, though let down by poorly executed observational drawings and diagrams.

Rackham, Thomas, *The Moon in Focus*, Pergamon Press, 1968. 183pp. Includes a good description of the lunar surface.

Sheehan, William P., and Dobbins, Thomas A., *Epic Moon*, Willmann-Bell, 2001. 363pp. A history of telescopic lunar exploration. Recommended.

Whitaker, Ewen A., *Mapping and Naming the Moon*, Cambridge University Press, 1999. 242pp. A history of lunar cartography and nomenclature. Recommended.

Wlasuk, Peter T., *Observing the Moon*, Springer, 2000. 181pp. A reliable guide to lunar observing techniques, containing model observations and impressive images. CD-ROM included.

Lunar science

Kopal, Zdenek, *An Introduction to the Study of the Moon*, Reidel, 1966. 464pp. A comprehensive guide to lunar science in the pre-Apollo era.

Heiken, G. H., Vaniman, D. T., and French, B. M. (editors), *The Lunar Sourcebook: A User's Guide to the Moon*, Cambridge University Press, 1991. 736pp. Despite its subtitle, this book is a technical guide to the Moon's geology and the analyses of the Apollo rock samples.

Spudis, Paul D., *The Geology of Multi-Ring Impact Basins*, Cambridge University Press, 1993. 263pp. How the large lunar (and other planetary) basins were formed, written by a lunar geologist.

Spudis, Paul, *The Once and Future Moon*, Smithsonian Institution Press, 1996. 306pp. A comprehensive guide to the history of the Moon and the processes of lunar geology, written by a geologist.

Taylor, Stuart Ross, *Lunar Science: A Post-Apollo View*, Pergamon Press, 1975. 372pp. Scientific results and insights into lunar geology based on Apollo samples, written by a geologist.

Atlases and maps

de Callatay, Vincent, *Atlas of the Moon*, Macmillan, 1964. 160pp. A good selection of close-up telescopic photographs of the terminator, with annotated line drawings and descriptions.

Chong, S. M., Lim, Albert C. H., and Ang, P. S. *Photographic Atlas of the Moon*, Cambridge University Press, 2002. 145pp. Day-by-day photographic coverage of the whole Moon throughout the lunation.

Elger's Map of the Moon (revised by H. P. Wilkins), George Philip, 1957. A clear black-and-white line drawing of the whole lunar disk, made to a diameter of 450 mm, with a short description of features.

Hatfield, Henry, *The Hatfield Photographic Lunar Atlas*, Springer, 1998. 130pp. A collection of close-up lunar photographs taken between 1965 and 1967, showing areas under different angles of illumination. Still very useful, even though modern CCD images show far more detail.

Kopal, Zdenek, *A New Photographic Atlas of the Moon*, Robert Hale, 1971. 311pp. Close-up lunar images taken with a variety of instruments and from various spaceprobes. More a tour of some of the finest features rather than a true atlas.

Philip's Moon Map, Philip's, 2003. The Moon's nearside clearly drawn, labelled and indexed, showing more than 500 features, with a text on how to observe the Moon and a guide to the best examples of various features. 56 cm in diameter.

Rükl, Antonin, *Hamlyn Atlas of the Moon*, Hamlyn, 1990. 224pp. A wonderfully clear, detailed atlas of the Moon in 76 sections, showing most objects visible through a 100 mm aperture. Thoroughly recommended.

Westfall, John E., *Atlas of the Lunar Terminator*, Cambridge University Press, 2000. 292pp. A large selection of CCD images showing features along the terminator through the lunar month, but the majority of the images are over-processed and too contrasty to be of much use in seeing fine tonal detail.

Globes

NASA Moon globe (Replogle)
A 300 mm diameter lunar globe showing the main features on a detailed light grey map.

Räth's Erdmondglobus (Rathgloben)
A beautifully drawn 330 mm diameter lunar globe.

Software

Lunar Map Pro
Publisher RITI
Website http://www.riti.com
An electronic Moon map for Windows 98/ME/NT/2000/XP, zoomable to high magnifications, with two modes – a detailed shaded map and an accurate vector graphic line rendition.

LunarPhase Pro
Publisher NovaSoft Ltd
Website http://www.nightskyobserver.com
A complete Moon utility for Windows 98/ME/NT/2000/XP with a wide range of capabilities. Recommended.

Virtual Moon Atlas
by Patrick Chevalley and Christian Legrand
Website http://www.astrosurf.com/avl/UK_index.html
This freeware package for Windows 95/98/NT/2000/XP is a must for the lunar observer who has a computer and wishes to plan observations. Packed with features and very easy to use. Recommended.

GENERAL INDEX

FEATURES INDEX

──── *ACKNOWLEDGEMENTS* ────

Thanks to Robin Scagell, who recommended me to Philip's. Frances Adlington, Commissioning Editor for Astronomy at Philip's, was immensely helpful and encouraging, and patiently saw the project through from beginning to end. Thanks to John Woodruff for his excellent editorial work in the book's final stages. Caroline Rayner has performed a super job in editing this revised edition.

The lunar studies of Mike Brown, Doug Daniels, Bill Dembowsky, Colin Ebdon, Harold Hill, Brian Jeffrey, Nigel Longshaw, Cliff Meredith, Bob Paterson, Paul Stephens, Grahame Wheatley, and their kind correspondence through the years, have provided much inspiration. Both Mike and Brian helped me get to grips with lunar CCD imaging, and I can now fully appreciate their skill in recording fine lunar detail electronically. The beautiful observational drawings of Colin, Harold, Nigel, Bob and Grahame demonstrate that sketching lunar features is by no means outmoded or inconsequential – it remains the best way of learning the Moon's topography. Paul's friendship, help and advice have been invaluable, and access to his superb private astronomical library has allowed me to consult many rare astronomical books. Jon Harper provided me with some very useful information on lunar occultations, and Peter Van Buren's comments were especially helpful.

Finally, I could never be a lunar observer – let alone have contemplated writing this book – without the patience and understanding of my wife, Tina, and my little daughter, Jacy, whose nocturnal slumbers I have often disturbed in the pursuit of the Moon's wonders.

Picture credits

Mike Brown 13, 17, 22, 24, 26, 29, 31, 67, 89, 90, 101, 103, 108; Doug Daniels 19, 97; Colin Ebdon 81; Peter Grego 1, 9, 16, 36, 41, 48, 51, 53, 54, 55, 76, 77, 85, 99, 113, 121, 122, 135, 140, 141, 143, 144, 146, 148, 149, 155, 159, 162, 175; Mike Goodall 127; Harold Hill 111; Brian Jeffrey 15, 112, 116; Nigel Longshaw 96; Cliff Meredith 66; Phil Morgan 71; NASA 38, 163, 165, 166, 167, 169, 170, 172, 176; Bob Paterson 121; Paul Stephens 5, 65, 79, 141, 158, 161; Grahame Wheatley 70, 78.

Moon map by John Murray © Philip's
Artwork © Philip's